Advance Praise for Plant-Forward Cuisine

"Ole and Klavs—a scientist and a chef—have beautifully combined their years of expertise to show how this amazing fusion of knowledge can help us to face the future by looking to tradition. This is a book to think about, to feel, to taste, and to enjoy..."
  Juan Carlos Arboleya, Professor at the Basque Culinary Institute

"Ole has done it again! A visionary book that re-imagines the future of sustainable food by centering on deliciousness. A recipe for a better, more pleasurable future. Read it. Share it. Read it again."
  Rob Dunn, Professor and Author of *Delicious*

"Ole's team challenge us to use umami, mouthfeel and our five senses to increase the variety of plant-based foods consumed, inspiring scientists, chefs and the curious home cook to reduce environmental impact."
  Sally Gras, Professor and Deputy Director of Plants4Space, Australia

"Another outstanding collaboration from physicist-gourmand Ole Mouritsen and chef Klavs Styrbæk, Plant-Forward Cuisine is a richly detailed guide to good eating in the 21st century: good for our own well being, and for the planet's."
  Harold McGee, Author of *On Food and Cooking* and *Nose Dive*

"In this book Ole and Klavs, with each their respective areas of expertise, give a necessary boost to some of the more overlooked members of the plant kingdom. Everything is underscored by solid research and a deep understanding of the science behind flavours, aromas, and textures. A wonderful read for chefs and amateur cooks alike."
  Rasmus Munk, Head Chef and Co-Owner, Restaurant Alchemist**

"A fascinating, bold, and beautiful journey into the science and deliciousness of plants."
  Amy Rowat, Professor and Founder of Science & Food, UCLA

"I can't decide what's more thrilling, Ole Mouritsen's astonishing depth of knowledge or Klavs Styrbæk's wildly creative recipes. Plant-Forward Cuisine is as delightful to read as it is beautiful to look at."
  Mark Schatzker, Author of *The Dorito Effect* and *The End of Craving*

"At last, an inspiring book, written by a chef and a scientist, on the future of eating that addresses the solution as well as the problem of how to eat healthily and sustainably for pleasure and for the planet."
  Barry Smith, Professor and Director of the U-London Center for the Study of the Senses

"Plant-Forward Cuisine is a culinary adventure for your taste buds and your mind. The dazzling photos and mouth-watering recipes will tempt you to get cooking and learning about the science of flavour, texture, and a sustainable future for our planet."
  Pia Sörensen, Co-author of *Science & Cooking*, Harvard University

Plant-Forward Cuisine

Plant-Forward Cuisine is a beautifully illustrated book that promotes the environmental and health benefits of a plant-forward diet and will inspire readers with a range of exciting recipes.

The book addresses the urgent need to make changes to those culinary cultures where animal-sourced proteins play a central role. To ensure that there is enough food for a growing world population, to lessen the burden on the environment, and to promote healthier, sustainable eating patterns, it is crucial to transition to a diet that focuses primarily on plants as the key ingredients. Yet, many people dislike the taste of plants because of their texture and lack of sweetness and umami. Luckily, the book provides a solution to these challenges. It offers key scientific descriptions of the physical characteristics of plants, mushrooms, algae, and fungi and their nutritional components, along with information about creation of texture and flavour. Armed with this knowledge, the recipes then provide tips and tricks for transforming plants into delicious meals with pleasing textures and flavours. The authors stress that it is not necessary to embrace a fully vegetarian or vegan diet. Rather, they suggest that taking a flexitarian approach, which incorporates small quantities of animal products to elicit umami, may be a more viable and lasting solution for people at large. Throughout the text readers will find interesting narratives about various aspects of green gastronomy around the world. The book concludes with two helpful reference sections: a glossary of main ingredients used in the recipes and a compilation of ingredients that can be used to add taste and aroma.

This book will be of great interest to those concerned with building a sustainable food system, and it will serve as a practical guide for those seeking to transition to plant-rich diets without compromising their taste experience and enjoyment of food.

Ole G. Mouritsen is a research scientist and Emeritus Professor of gastrophysics and culinary food innovation at the University of Copenhagen, Denmark. He is the President of the Danish Gastronomical Academy and Founder and past Director of the National Danish Taste Centre, Taste for Life.

Klavs Styrbæk is a professional chef who owns and operates STYRBÆKS. He is the author of the award-winning cookbook *Mormors Mad* (Grandmother's Food, 2006). He has twice been awarded an honours diploma for excellence in the culinary arts by the Danish Gastronomical Academy.

Mariela Johansen, who has Danish roots and an international background, lives in Canada. She has worked with Ole and Klavs to translate and adapt several of their books for an English language readership, two of which have won regional Gourmand World Cookbook Awards in the translation category.

ROUTLEDGE STUDIES IN FOOD, SOCIETY AND THE ENVIRONMENT

Holiday Hunger in the UK
Local Responses to Childhood Food Insecurity
Michael A. Long, Margaret Anne Defeyter and Paul B. Stretesky

School Farms
Feeding and Educating Children
Edited by Alshimaa Aboelmakarem Farag, Samaa Badawi, Gurpinder Lalli and Maya Kamareddine

The Vegan Evolution
Transforming Diets and Agriculture
Gregory F. Tague

Food Loss and Waste Policy
From Theory to Practice
Edited by Simone Busetti and Noemi Pace

Rewilding Food and the Self
Critical Conversations from Europe
Edited by Tristan Fournier and Sébastien Dalgalarrondo

Critical Mapping for Sustainable Food Design
Food Security, Equity, and Justice
Audrey G. Bennett and Jennifer A. Vokoun

Community Food Initiatives
A Critical Reparative Approach
Edited by Oona Morrow, Esther Veen and Stefan Wahlen

Food Futures in Education and Society
Edited by Gurpinder Singh Lalli, Angela Turner and Marion Rutland

The Soybean Through World History
Lessons for Sustainable Agrofood Systems
Matilda Baraibar Norberg and Lisa Deutsch

Urban Expansion and Food Security in New Zealand
The Collapse of Local Horticulture
Benjamin Felix Richardson

Evaluating Sustainable Food System Innovations
A Global Toolkit for Cities
Edited by Élodie Valette, Alison Blay-Palmer, Beatrice Intoppa, Amanda Di Battista, Ophélie Roudelle and Géraldine Chaboud

How to Create a Sustainable Food Industry
A Practical Guide to Perfect Food
Melissa Barrett, Massimo Marino, Francesca Brkic and Carlo Alberto Pratesi

Food Justice in American Cities
Stories of Health and Resilience
Sabine O'Hara

Sustainable Food Procurement
Legal, Social and Organisational Challenges
Edited by Mark Stein, Maurizio Mariani, Roberto Caranta and Yiannis Polychronakis

American Farming Culture and the History of Technology
Joshua T. Brinkman

Transforming Food Systems
Narratives of Power
Molly D. Anderson

The Real Cost of Cheap Food
Michael Carolan

Plant-Forward Cuisine
Basic Concepts and Practical Applications
Ole G. Mouritsen, Klavs Styrbæk and Mariela Johansen

For more information about this series, please visit:
www.routledge.com/Routledge-Studies-in-Food-Society-and-the-Environment/book-series/RSFSE

PLANT-FORWARD CUISINE

Basic Concepts and Practical Applications

Ole G. Mouritsen
Klavs Styrbæk
Mariela Johansen

Photography and design
Jonas Drotner Mouritsen

LONDON AND NEW YORK

Designed cover image: © Getty Images

First published in English 2025 by Routledge
4 Park Square, Milton Park, Abingdon, Oxon OX14 4RN

and by Routledge
605 Third Avenue, New York, NY 10158

Published in Danish by Gyldendal A/S 2020

Routledge is an imprint of the Taylor & Francis Group, an informa business

© 2025 Ole. G. Mouritsen, Klavs Styrbæk, and Mariela Johansen

The right of Ole. G. Mouritsen, Klavs Styrbæk, and Mariela Johansen to be identified as authors of this work has been asserted in accordance with sections 77 and 78 of the Copyright, Designs and Patents Act 1988.

All rights reserved. No part of this book may be reprinted or reproduced or utilised in any form or by any electronic, mechanical, or other means, now known or hereafter invented, including photocopying and recording, or in any information storage or retrieval system, without permission in writing from the publishers.

Trademark notice: Product or corporate names may be trademarks or registered trademarks and are used only for identification and explanation without intent to infringe.

British Library Cataloguing-in-Publication Data
A catalogue record for this book is available from the British Library

Library of Congress Cataloging-in-Publication Data
Names: Mouritsen, Ole G., author. | Styrbæk, Klavs, author. | Johansen, Mariela, author. | Mouritsen, Jonas Drotner, photographer.
Title: Plant-forward cuisine : basic concepts and practical applications / Ole G. Mouritsen, Klavs Styrbæk, Mariela Johansen ; photography and design Jonas Drotner Mouritsen.
Description: London ; New York : Routledge, Taylor & Francis Group, 2025. | Includes bibliographical references and index. | Identifiers: LCCN 2024030121 (print) | LCCN 2024030122 (ebook) | ISBN 9781032765396 (hardback) | ISBN 9781032765372 (paperback) | ISBN 9781003478959 (ebook)
Subjects: LCSH: Vegan cooking. | Food--Analysis. | LCGFT: Cookbooks.
Classification: LCC TX837 .M758 2025 (print) | LCC TX837 (ebook) | DDC 641.5/6362--dc23/eng/20240907
LC record available at https://lccn.loc.gov/2024030121
LC ebook record available at https://lccn.loc.gov/2024030122

ISBN: 978-1-032-76539-6 (hbk)
ISBN: 978-1-032-76537-2 (pbk)
ISBN: 978-1-003-47895-9 (ebk)
DOI: 10.4324 / 9781003478959

Publisher's Note
This book has been prepared from camera-ready copy provided by the authors

Typeset in Turnip by Chromascope.dk

Printed and bound in Great Britain by
TJ Books, Padstow, Cornwall

Contents

Recipes	viii
Gastonomic detours	ix
The people behind the book	xi
Preface	xiii

Green Cuisine — 1

Toward a plant-forward diet — 5
The question of taste—meat versus plants — 5
The question of texture—is the food crisp enough? — 7

Toward a planet-friendly diet — 11
Food in the Anthropocene epoch — 12
Change is needed — 12
A model planetary diet — 13
Understanding resistance to change — 13
Attitudes are evolving — 15

The building blocks of green cuisine — 17
In the ground and along its surface —root vegetables — 18
Above the ground—greens, leaves, and herbs — 18
Flowers — 28
Fruits, legumes, and pulses — 28
Nuts and small seeds — 28
Edible mushrooms and other fungi — 29
Greens from the deep blue sea: microalgae and seaweeds (macroalgae) — 29
Very green and nutritious, but forget about taste and texture — 30
Seaweeds—an integral part of the food web — 31

Understanding the taste experience — 41
The basics of taste — 41
The taste of plants — 44
The taste of mushrooms — 60
Seaweeds—the taste and smell of the sea — 66

Green cuisine comes in many colours — 71
Four types of pigments — 71
The colours of mushrooms — 74
Colourful seaweeds — 78

Adding taste and texture — 81
The Flavour Accelerators — 82
How to meet the taste challenge head-on — 83
Texture needs special care — 84
The structure and function of plant cells — 84
The effect of heat on texture — 88

The nutritional underpinnings of green cuisine — 97
Food and good health — 97
Proteins — 99
Fats — 104
Carbohydrates — 106
Dietary fibre — 112
Antioxidants — 115
Vitamins — 115
Micronutrients—minerals and trace elements — 118
Natural poisons in plants and mushrooms — 119
Raw, never cooked — 120
Plant-forward ingredients and food hygiene — 121
Preserving and fermenting — 121

Organic and conventional farming — 125
Is organic produce really better? — 125
Can enough food be grown using organic methods? — 126
A convergence of organic and conventional farming may be underway — 127

Final words — 128

A glossary of the main plant ingredients from A to Z — 131

The Flavour Accelerators — 147

Recipes — 171

Bibliography	272
Illustration credits	275
Index	276
Acknowledgements	284

Recipes

The recipes are for appetisers, entrées, and side dishes, as well as seasonings listed under *The Flavour Accelerators*. In some cases the number of servings should be adjusted taking into account how the dish fits into the overall meal. As well, one can prepare some of the Flavour Accelerators in quantities greater than those for a single use. They can then be stored as indicated in the text to be within easy reach another time.

The flavour accelerators	**147**
Aïoli	147
Anchovy sauce (*bagna càuda*)	148
Basic vegetable bouillon	150
Citrus dressing	151
Dashi	152
Fruit vinegar marinade	153
Meligarum	154
Gastrique	154
Pickled ginger (*gari*)	154
Gremolata	155
Hemp seed crunch	155
Hoisin sauce	156
Lime-soy-fish sauce dressing	157
Miso mayonnaise	158
Dressing with nutritional yeast	159
Kampot pepper sauce	161
Pink peppercorn sauce	161
Seaweed pesto	161
Horseradish-parsley pesto with capers	161
Ponzu and *sanbaizu*	162
Seeds with soy sauce	163
Tamarind cashew nuts	163
Salted, smoked peanuts	163
Romesco sauce	163
'Seaweed liquorice'	165
Simmered tomato sauce with herbs and vegetables	167
Basic *tsukemono* marinade	167
Za'atar	169
Pickled and marinated	**172**
Cucumber *tsukemono*	172
Kohlrabi *tsukemono*	173
Salt-pickled lemons	174
Salt-pickled beans and plums with savory	176
Koji-marinated broccolini with chilli sauce	178
Kimchi	179
Salads	**180**
Koji-marinated vegetables	180
Kidney bean and *tsukemono* salad	180
Celeriac salad	182
Radicchio	184
Wakame salad with tofu	184
Cucumbers with sesame seeds, *ponzu*, chilli, and ginger	186
Soups	**188**
Miso soup with seaweed	188
Pea soup with scallops	188
Onion soup—the real thing!	190
Dishes with eggs	**192**
Sautéed spinach, mung beans, and wood ear mushrooms	192
Rizza	194
Kale pancakes	196
Tortilla and figs	198
Fried eggs on a bed of celeriac	200
Vegetable stews	**202**
'Greens to go' casserole	202
Green fricassee	204
Sicilian ratatouille	206
Mashed vegetables with rutabaga 'bacon'	208

Aubergine dishes	210	**Potatoes**	242
Deep-fried aubergines *(nasu dengaku)* with *miso*	210	Potato purée	242
Juicy aubergines	212	Potatoes with a difference	244
Aubergines *au gratin*	214	Potatoes with fresh cheese and raw mushrooms	246
Aubergine 'fettucine'	216	Pears, beans, potatoes, and smoked pork belly	248
Green 'lasagna'	218	Sweet potatoes	250

Aubergine dishes 210
Deep-fried aubergines *(nasu dengaku)* with *miso* 210
Juicy aubergines 212
Aubergines *au gratin* 214
Aubergine 'fettucine' 216
Green 'lasagna' 218

Asparagus dishes 220
Green asparagus 220

Brassica family 222
Broccoli with *miso*-mayonnaise 222
Crisp cabbage 224
Cone cabbage with asparagus 226

Root vegetables 228
Dry-cured carrots *à la gravlax* 228
Carrots in zucchini sauce 228
Baked root vegetables 230
Baked sunchokes 232
Simmered *daikon* 234
Red beets and kidney beans 236

Onions and leeks 238
Grilled onions 238
Succulent leeks 240

Potatoes 242
Potato purée 242
Potatoes with a difference 244
Potatoes with fresh cheese and raw mushrooms 246
Pears, beans, potatoes, and smoked pork belly 248
Sweet potatoes 250

Grains 252
Couscous 252
Quinoa salad with dulse 254
Quinoa, cauliflower, and tomatoes 255
Polenta fritters with blue cheese dip 256
Baked squash with mushroom risotto
    and crumble 258
Pasta *tarako* 260

Mushrooms 262
Mushroom pâté 262

Beans and pulses 263
Yellow split pea hummus with thyme 263
Lentils, romanesco, and pomelo 264
White beans with seaweeds 266

Fruits 268
Tomatoes in tomatoes, with mouthfeel 268
Chilli-watermelon 270

Gastonomic detours

A Catalan celebration of a special onion 20
Japanese heirloom vegetables—a specialty of the
    Kyoto Prefecture 24
Shōjin ryōri—taking green ingredients
    to a higher level 32
Bananas and their rhizomes—a cautionary tale 38
True tastes and 'taste-like sensations' 43
Koku attributes and *kokumi* 47
Ajinomoto—where umami and *kokumi*
    seasonings were launched 48
The Dorito effect—flavour based on additives 52
Tomatoes are not like they used to be 52

Chilli peppers—from the mild
    to the smoking hot 56
Truffles—in a category by themselves 62
Browning 76
Why do some vegetables become woody? 90
Tsukemono—creating umami and
    interesting textures the Japanese way 92
Koji has an almost magical effect on vegetables 95
Proteins and the transformation of the
    Japanese diet 102
The potato—a somewhat neglected vegetable 108
Tsukemono—a nutrient powerhouse 123

Shortly before this book went into print, Mariela Johansen, our devoted co-author and collaborator through fifteen years, passed away.

We wish to dedicate the book to the memory of Mariela.

Klavs & Ole

The people behind the book

Ole G. Mouritsen is a research scientist and professor emeritus of gastrophysics and culinary food innovation at Copenhagen University. His work in the past decade has focused on basic sciences and their applications within the fields of biophysics, biomedicine, and food. He is the recipient of numerous prizes for his work and for research communication. His extensive list of publications includes four monographs co-authored with Klavs Styrbæk, which integrate scientific insights with culinary perspectives and have been nominated three times for Gourmand Best in the World Awards. Currently, Ole is president of The Danish Gastronomical Academy and past director and founder of the national Danish centre *Taste for Life*, a cross-disciplinary centre that aims to foster a better understanding of the fundamental nature of taste impressions and how we can use this knowledge to make much more informed and healthier food choices. Its extensive educational programme reaches out to audiences of all ages, with a special effort directed toward children to shape their dietary habits from an early age. For many years, Ole has been fascinated with the Japanese culinary arts and explaining the extent to which its techniques and taste elements can be adapted for the Western kitchen. In recognition of his efforts, he was appointed in 2016 as a Japanese Cuisine Goodwill Ambassador by the Japanese Ministry of Agriculture, Forestry, and Fisheries, and in 2017 the Japanese Emperor bestowed upon him The Order of the Rising Sun, Gold Rays with Neck Ribbon, Kyokujitsu chujusho 旭日中綬.

Klavs Styrbæk is a professional chef who owns and operates STYRBÆKS together with his wife, Pia. By combining a high standard of craftsmanship, sparked by curiosity-driven enthusiasm, he has created a gourmet centre where people can enjoy excellent food and where they can come to learn and take their culinary skills to a whole new level. Klavs is particularly enthusiastic about seeking out unique, local raw ingredients that are incorporated into new taste adventures or used to revisit traditional Danish recipes that might otherwise be forgotten. This delicate balance between innovation and renewal is demonstrated in his award-winning cookbook *Mormors Mad* (Grandmother's Food) (2006), which was honoured with a special jury prize at the Gourmand World Cookbook Awards in 2007. In 2008 and 2019 he was awarded an honorary diploma for excellence in the culinary arts by the Danish Gastronomical Academy. Many of the recipes that appear in the books co-authored with Ole originated in the test-kitchens at STYRBÆKS.

Mariela Johansen, who has Danish roots, lives in Vancouver, Canada. She holds a Master of Arts degree from Simon Fraser University, which focused on Humanities with a special interest in the ancient world. Working with Ole and Klavs, she has translated their four monographs, adapting them for a wider English language readership. Two of these, *Umami: Unlocking the Secrets of the Fifth Taste* and *Mouthfeel: How Texture Makes Taste*, won a Gourmand World Cookbook Award for the best translation of a cookbook published in the USA in 2014 and 2017, respectively.

Jonas Drotner Mouritsen is a graphic designer and owner of the design company Chromascope, which specialises in graphic design, animation, and film production. His movie projects have won several international awards. In addition, he has been responsible for layout, photography, and design of several books about food, some of which have been nominated for Gourmand World Cookbook Awards.

Preface

Many people will no doubt wonder whether there is a place for yet another volume that focuses primarily on plant-based recipes. Bookstores and libraries are already bursting with cookbooks covering the preparation of just about every type of edible plant and highlighting an impressive range of culinary traditions. So how can we make the case for inviting readers to join us on a food adventure to explore what we think of as 'green cuisine'—plant-forward, planet friendly, and based on an abundance of vegetables, along with some fruits, mushrooms, and seaweeds?

Part of the answer lies in our approach to the subject, based on our perspectives as a scientist and a chef, respectively. For many years Ole and Klavs have collaborated to investigate the physical properties and chemical make-up of raw food ingredients and to experiment with how best to prepare them in order to enhance their taste and texture. We have also benefited from feedback gained in the course of numerous outreach activities to groups of all ages, ranging from small children to the elderly.

It became clear to us that the preparation of the edible plants traditionally referred to as vegetables was a somewhat neglected culinary art in those food cultures, including our own, where they do not feature prominently. We, therefore, decided to integrate the knowledge that we had accumulated over the years into a book focusing specifically on the ingredients that are central to a plant-forward diet. We think that there are several reasons why it is both timely and necessary to demonstrate that these foods are not only nutritious and a vital component of a healthy diet, but also have tremendous gastronomic potential.

Our advocacy for green cuisine is rooted in serious concerns, shared by many others, about the negative aspects of the current human diet, which has evolved over the centuries to include a substantial proportion of animal protein. It contributes to climate change in a significant way, will not be able to provide sufficient nourishing food for a growing global population, and is associated with undesirable health outcomes. So, it is inevitable that many people will be faced in the short term with increasing pressures to make drastic changes to their ingrained habits. This will be a difficult adjustment, and some will turn to the meat substitutes that are now being promoted as a way around the problem, even though these are highly processed factory foods, generally based on grains, legumes, and peas. While their environmental footprint is lower than that associated with the production of conventional animal protein, it is not negligible. Using these ingredients in their natural state is much more efficient and remains by far the most economical and sustainable option. That is why this book stresses the importance of a scientific understanding of the structure and make-up of the raw ingredients, which determines how best to prepare them. Our goal is to build on that basic knowledge to inspire readers with recipes for delicious plant-based dishes to which they can turn as a conscious choice before it becomes a matter of necessity.

In earlier books we have written about seaweeds, umami, mouthfeel, cephalopods, and Japanese pickles. We have incorporated some of the ideas that are explored in greater depth in these volumes into this book, as well as a few of their recipes. We have also drawn on related scientific articles and found inspiration in publications about food intended for a general audience. These references are listed in the bibliography at the back of the book for readers who wish to delve deeper into these topics.

Mariela has translated Ole's and Klavs' four earlier books and adapted them for publication in English. Two of them, *Umami: Unlocking the Secrets of the Fifth Taste* (2014) and *Mouthfeel: How Texture Makes Taste* (2017), were nominated for a Gourmand World Cookbook Award. She has written the English version of *Plant-Forward Cuisine* based on her own translation and revision of the original 2020 Danish-language edition. By restructuring the text to frame it more clearly in today's context and inserting interesting historical details and contemporary references, Mariela has reworked the book for an international readership.

It would not have been possible to produce a book of this scope without the assistance and support of many others and we are deeply grateful for their help and advice. We extend our thanks to them in the acknowledgements.

Ole G. Mouritsen, Klavs Styrbæk, and Mariela Johansen

GREEN CUISINE

a healthy, environmentally sound culinary approach

In the past few years there has been a growing consciousness of the close relationship between food production and consumption and the state of the planet. Environmentalists, concerned about the effects of climate change, are advocating for a plant-rich diet in order to arrive at more ecologically sound, sustainable, and equitable ways to utilise the Earth's food resources. Famous chefs and proponents of vegetarian and vegan lifestyles all have their particular take on replacing some or all of the animal products in our daily diet. Nutritionists have warned that the rise in the incidence of cardio-vascular diseases, diabetes, and obesity in some parts of the world is, at least in part, attributable to people's eating habits. So, we are hardly the first to suggest the need for a fresh approach in the kitchen, a green cuisine that is plant-forward and planet friendly.

Based on our outreach activities, we know only too well that fruits of all kinds are well-liked, but there are many who are unenthusiastic about eating vegetables and even outright reject certain ones. This is memorably illustrated by the late American President George H. W. Bush's defiant declaration: "I do not like broccoli. And I haven't liked it since I was a little kid and my mother made me eat it. And I'm President of the United States and I'm not going to eat any more broccoli." We have learned, however, that even unequivocal statements of this kind should not discourage us. There are many simple techniques for turning plant-forward food of virtually any kind into such appetising dishes that all, including children, are happy to eat them and become much more willing to try new and different ingredients.

Our ideas are to a large extent based on three main principles that we have modelled on Asian food cultures, with traditions that go back at least a thousand years. These food cultures are justly renowned for being able to prepare delicious and healthy meals using vegetables, herbs, seaweeds, and mushrooms. In many cases, for example, the Zen temple cuisine in Japan, these ingredients are used exclusively. One of the secrets of this style of cooking is to find ways to bring the fifth taste, umami, into the picture using seasonings and condiments that complement the taste of the raw ingredients themselves. Another is to ensure that the dishes have an interesting mouthfeel, with crisp and crunchy textures being especially important. Finally, it is important that the food engages all five senses. It needs to do more than appeal to the taste buds—it should feel good on the tongue, possibly make an interesting sound when we bite into it, give off a distinctive aroma, and look appetising. Eating plant-forward food should be a multisensory experience.

Even though this book focuses on helping people to move toward a more plant-based diet, we do not see the need to exclude any food groups and are not dogmatic about it, taking a somewhat flexitarian approach. Small quantities of meat, fish, shellfish, molluscs, and roe can be used to great effect in a supporting role to make a dish more appealing to those who expect to find them on their plates. Animal products, such as cheese, yogurt, and eggs, and extracts, powders, and fermented pastes made, for example, from insects, fish, shellfish, and cephalopods, are also of great value.

And we advocate for the use of microorganisms, fungi, and enzymes to create interesting taste nuances and add enticing aromas to those vegetables that might otherwise be a bit bland. Last, but not least, we also make use of a range of spices, herbs, and seasonings that elicit other flavours or add texture. We think of all of these collectively as *The Flavour Accelerators*—ingredients that can be kept on hand or prepared ahead of time, have good keeping qualities, and are readily available to give a dish a bit of a lift.

At the beginning of the book, we home in on the questions that are central to our approach to green cuisine. What is it about mankind's evolutionary path that leads so many people to dislike the taste of plant-based dishes? And how is this related to the biology of plants? What can be done to address the increasingly urgent problem of finding a sustainable way to provide a healthy diet for a growing global population? In addition, there is a discussion of research into the factors that influence our taste preferences, especially those of children and young people, and our resistance to change. The answers to these questions may encourage those who have not yet turned to a plant-forward diet to consider the benefits of adopting some or all of its aspects.

In order to lay the groundwork for the recipes that follow, we first look at the physical structure of plants, mushrooms, and seaweeds. Because of the particular way in which these living organisms grow they are often more difficult to prepare and may be less palatable than meat or seafood. We then describe those which we have included in the book under the umbrella of green cuisine. This leads to a discussion of how we can characterise our gustatory perception of them, taking into account that all five senses are involved when we evaluate what we think of as the 'taste' of plant-based food. And naturally, we also touch on the colours of the raw ingredients—paradoxically, even though they are used in green cuisine, many of them are not green.

Next we present a general overview of the potentially beneficial effects of following a plant-forward diet, as well as the special challenges associated with adhering to a vegetarian or a vegan diet. This is supplemented by detailed information about the nutritional elements found in a selection of plants, mushrooms, and seaweeds, and advice regarding safe ways to store and preserve these raw ingredients.

Finally, we take a brief look at some aspects of organic and conventional farming practices and whether we should take these into consideration when choosing raw ingredients.

Armed with a basic understanding of the inherent properties of ingredients that are used in plant-forward dishes, we come to the second core part of the book—the recipes and suggestions that will help us to transform them into delicious food. To do so, we will present many tips and tricks for preparing them, including making use of *The Flavour Accelerators* to add umami and other enticing flavours. In addition, we will explain how one can take advantage of the structure of vegetables to enhance their mouthfeel, especially so that they are crisp and succulent. Having a good grasp of these concepts will allow readers to improvise, to become more creative, and to develop dishes of their own.

The book concludes with two handy reference sections: a glossary of the raw ingredients used in the recipes, as well as their characteristic aromas and tastes and how they are affected by various preparation methods, and a comprehensive inventory of *The Flavour Accelerators* and their applications.

The American humorist Will Rogers once said that "An onion can make people cry but there has never been a vegetable invented to make them laugh." This is true enough. But with the help of this book, readers may discover ways to prepare delicious plant-based dishes that will bring a smile to their face.

TOWARD A PLANT-FORWARD DIET

humans are not naturally herbivores

In many food cultures, vegetables are relegated to second place in the culinary hierarchy, ranking well below meats, fish, and sauces. When asked about why they may prefer meat to vegetables, many people in Western food cultures mention that undesirable 'tastes' prevent them from enjoying plant-based dishes. These dishes often pose challenges to our sensory perception, whether these relate either to the raw ingredients themselves or to how they are prepared. They primarily have to do with taste and texture and, possibly, result from some preconceived ideas. How do we feel about vegetables in general? Do we like them? Have we even tasted some of the less well-known ones? This leads directly to the question of whether we can do anything to alter their taste and texture so that they are more to our liking and so that we might be more inclined to eat them. Can we bring the inherent deliciousness and aromas of the plants to the forefront, or can we enhance them with seasonings? And are there ways to alter the texture of the raw ingredients to turn them into better and more interesting eating experiences, both with respect to how they feel on the palate and what it is like to chew on them?

There are very distinct answers to these questions that can be found by examining human food preferences and the role of plants in our diet. These are a function, respectively, of something as fundamental as our evolution as a species and of the biology, as well as the actual physical and chemical structure, of plants.

The question of taste—meat versus plants

The food preferences that now disadvantage vegetables can be linked to our evolutionary history. Broadly speaking, scientists regard the period about 1.9–1.8 million years ago as a time of transition from the ape species, *Australopithecines*, to the hominins, *Homo erectus*, who had a smaller brain and a lower brow than modern humans. It is presumed that an important driving force for this evolution was that, during this transition period, the ancestors of *Homo erectus*, the habilines, became hunter-gatherers and changed from herbivores into omnivores. Although our own species, *Homo sapiens*, is thought to date back only about 200,000 years, our forebears appear to have been eating meat for at least two million years.

A Harvard anthropologist, Richard Wrangham, has proposed that the evolution of our early ancestors was linked to their having learned to use fire to cook food about 1.9 million years ago. Humans are the only living beings who prepare some of their food by heating it. The other large primates eat only raw food, and in order to derive enough energy from it, they have to spend about eight hours a day to gather it and an equal amount of time chewing it. Cooking plants and meat over a fire made the food more digestible and resulted in more readily available nutritional substances and, hence, a greater caloric intake. Cooking gelatinates the carbohydrates in the starchy parts of plants and denatures proteins in meats. As a consequence, humans need much less time to eat and digest their food than other animals. The hypothesis that early hominins started to cook much earlier than at first supposed is supported by how the facial characteristics of humans have evolved. We now have a smaller mouth and teeth and weaker jaws than other higher primates. We also have a smaller stomach and shorter intestines. This seems to be an indication that our digestive system has adapted to prepared food, consumed in smaller pieces that could be converted in less time into more nutrients and more energy. Wrangham views these evolutionary changes as evidence that humans had become cooks.

The plant-eating herbivores from the time before meat was introduced into our diet most probably lived in trees and derived the bulk of their nutrition from fruits and berries. The aromatic substances emanating from these kinds of plants would have been, as they still are, an indication that they were ripe, sweet, nutritious, and rich in calories. Once these ancestors started to eat meat, especially after they discovered how to use fire to cook their food, their diet became a much better source of proteins and calories. It is said that this created the conditions for the evolution of the large human brain. The combination of the sweet tastes from ripe fruits and of savoury umami from cooked meat became deeply embedded early on and these have, to a large extent, driven our food preferences for many millennia. Furthermore, human breast milk is characterised by these two tastes. As a consequence, babies develop a preference for sweet, savoury food as a basic condition of life, and this more or less leaves out vegetables.

The biology of plants also tends to work against their becoming a much more important component of our daily diet. Unlike animals, plants have no muscles and cannot move around. They are not meaty and do not store up large quantities of ATP (adenosine triphosphate), which is the energy source that fuels motion in animals. When we eat meat, its ATP content is broken down to create substances that have an intense umami taste. Given this lack of mobility, plants in the wild are unable to escape from animals that want to eat them, and some have developed a number of chemical defence mechanisms as a matter of self-preservation. They may either contain, or have ways to produce, substances that are poisonous or very bitter. Only those

Toward a plant-forward diet

ripe fruits that taste sweet and release attractive aromatics, are designed specifically to be eaten by animals. This helps to scatter their seeds to new areas where they can continue to grow, thereby propagating the species. In other cases, grasses and leaves form the mainstay of many animal diets, but while those plants are edible, they do not depend on being consumed in order to survive.

As the vast majority of plants, in particular the ones we think of as vegetables, have almost no natural sweet and umami tastes, they are intrinsically less attractive to humans. Unripe fruiting bodies are usually hard and sour, green leaves and stems can be bitter, parts of the plants might be poisonous, and edible roots are often quite hard. Nevertheless, virtually all humans continue to incorporate plants into their diet at some point or another. Fortunately, in the course of hundreds of years, people have discovered selective breeding in order to turn many of the original wild plants into crops that have greater nutritional value and are more palatable, easier to prepare, and relatively free of toxic substances.

Still, most people are not vegetarians, and the taste and texture of plants may discourage the majority of us from loading up on plant-based food every day. This is a natural tendency given the fundamental biological fact that humans have evolved to be meat-eaters. This is where the culinary arts and gastronomy come in to try to resolve this conundrum. By gaining an understanding of what we most enjoy about the taste of meat we can capitalise on it to make vegetables more appealing. We can also put our experience as cooks to work to turn out nutritious, filling green cuisine. And herbs, mushrooms, and seaweeds add diversity to our food sources that far surpass those we find in the animal kingdom. Finally, one might add that meat and meat dishes are rarely as attractive to look at and as aesthetically pleasing as a serving of vegetables that incorporates a variety of colours, shapes, textures, and aromas—a true multisensory food experience.

Yet another way to promote the adoption of a more plant-based diet is by pairing the food with appropriate beverages. The lack of umami in the vegetable dishes can, to some extent, be compensated for by choosing drinks that have it in abundance, especially fermented beverages such as aged wines, sake, beer, and kombucha, many of which are available in non-alcoholic versions.

Bitterness is not the same for everyone

One issue that affects the way that different people react to the taste of vegetables is related to perceptions of bitterness. Everyone has two copies of a special gene (TAS2R38) that codes for the taste receptors in the tongue that identify bitter substances. There are two variants of this gene (AVI and PAV). People who have two AVI genes are not sensitive to bitter tastes, while those who have an AVI and a PAV can taste them, and those with two PAV genes are particularly sensitive to bitterness. At one time those in the last group were called 'supertasters,' but this is somewhat misleading. Recent research has shown that people who carry the PAV gene are much less likely to choose bitter vegetables such as broccoli and Brussels sprouts, but their sensitivity to salty and sweet tastes is the same.

The question of texture—is the food crisp enough?

Green cuisine is not challenging solely because of how plants taste—there is also a question of texture. Our preferences are to a large extent driven by how food feels in our mouth and between our teeth when we chew on it. A lot of people cannot stand soggy vegetables or hard seeds. Most prefer their vegetables to be crisp and succulent.

Expressions such as hard, soft, tough, tender, dry, creamy, crispy, mealy, woody, sticky, and crunchy describe textures rather than tastes, but are nevertheless central to our overall impression of a particular food, especially a plant-based one. This, in many cases, is what determines whether we will eat a given ingredient, whether raw or prepared.

Texture, also known as mouthfeel, is defined as that aspect of the physical structure of a food that we can feel. It is a tactile sensory impression which is registered by especially sensitive nerve endings on our tongue, in our teeth, and in the oral cavity. These nerve endings are similar to those in many places on our skin and in our body that can detect pressure, vibrations, temperature, and pain.

The structure and, hence, the texture of plants, mushrooms, and to a certain extent seaweeds are primarily determined by the conditions under which they live. Their cells, unlike those of animals, are protected by a cell wall built up from certain fibres, providing support and mechanical strength. In the case of plants, the cells together create larger structures such as stems and leaves that are sufficiently stiff to allow them to stay upright and spread out to catch the rays of the sun so that they can photosynthesise and set fruiting bodies.

Plants and some seaweeds rely mainly on cellulose to strengthen their cell walls. Cellulose is a carbohydrate, which is insoluble in water and which humans cannot digest because our digestive system does not contain enzymes that can break it down. That is why cellulose is classified as an insoluble dietary fibre.

When a plant cell has reached its optimal size, and to a greater degree as it ages, a special substance called lignin is deposited in spaces in the cell walls, leading to the sort of rigidity that we know from trees. Vegetables can also have varying amounts of lignin, usually associated with a 'woody' mouthfeel. Mushroom cell walls have no cellulose, but instead contain another insoluble carbohydrate, chitin. This, oddly enough, is the same material that makes up the hard exoskeleton of insects and crustaceans and that results in a slightly crunchy mouthfeel. Both cellulose and chitin ensure that vegetables and mushrooms remain somewhat firm when they are cooked.

Plants, mushrooms, and seaweeds depend on a whole range of carbohydrates that are very different from each other to bind their cells together. In contrast to cellulose and chitin, these are water soluble. For example, plants use pectin and hemicellulose, while seaweeds rely on alginate, agar, and carrageenan, and mushrooms have glucans. What these have in common is that they are very good at binding water. This is also why these carbohydrates can be used as thickening and gelation agents in the kitchen.

The mouthfeel of plants, mushrooms, and seaweeds is also largely determined by the different carbohydrates that they accumulate to make up their energy depots. Plants rely on starch and inulin, while seaweeds use laminarin and starch-like substances. The energy depots in mushrooms, which like animals do not carry out photosynthesis, are composed of fats and glycogen rather than carbohydrates. For this reason, their texture is primarily determined by the presence of chitin.

The energy that plants need to grow is mostly stored in the form of starch, for example, in potato tubers, corn kernels, and rice grains where it is packed closely, in an almost crystalline arrangement. When these types of raw ingredients are heated in a liquid, the starch crystals melt and absorb a great deal of the liquid, a process known as gelatinisation. This makes the food softer, but also drier and more mealy.

In later chapters, we will describe the underlying structure of various green ingredients in order to gain some insight into how they are affected by the way they are prepared in the kitchen. Armed with this knowledge, it is easier to meet the texture challenge head on, especially when it comes to ensuring that the vegetables remain crisp when that is desirable.

But first, we should digress to consider how the dual challenges posed by climate change and feeding a growing world population may cause us to move toward a plant-forward diet more as a matter of necessity than of choice.

TOWARD A PLANET-FRIENDLY DIET

a question of sustainability

We are now living in what has been labelled the Anthropocene epoch, that period in history when humans are making an irreversible impression on the environment. Our activities are having an enormous effect on the inanimate physical nature of the planet, its ecosystems, and above all the climate.

Earlier epochs were divided according to major geological events. The Anthropocene epoch is different and there is no real agreement about when it actually started. Nevertheless, many researchers have linked it to the adoption of agriculture as a defining development. From that point forward, the food supply and the systems in place to provide it have become important factors. There is no doubt that industrialisation, including the changes in farming practices, has had its own impact and that, especially during the past fifty to one hundred years, the pressure on the planet's resources due to human activities has accelerated rapidly, in fact, nearly exponentially.

These concerns are outlined in the 2030 Agenda for Sustainable Development, a shared blueprint for peace and prosperity for people and the planet, published by the United Nations in 2015. Of the seventeen goals outlined in that document, more than half are related directly or indirectly to food and wellness. These include poverty and hunger, good health and well-being, clean water, affordable energy, responsible consumption and production, the oceans and the land, climate action, and innovation.

In the last few years, it has become increasingly difficult to ignore the urgent need to work towards these goals. One way to do so involves revisiting the way in which our evolutionary path has led us to a diet that incorporates such a substantial proportion of animal products. Apart from having implications for our health and well-being, it is a significant contributor to climate change and is no longer considered environmentally defensible, even in the short term. And it may hinder our efforts to provide sufficient nutrition for a growing world population.

We think that embracing a plant-forward cuisine can help, at the level of our individual actions and the local choices we make, to address these global issues.

Food in the Anthropocene epoch

Only a minority of people still refuse to recognise that climate change is to a large extent anthropogenic, and most of us would agree that the way in which we have set up our food production systems plays an important role in the state of the environment. Even though more than 800 million people are starving and suffer from malnutrition, the total food supply has, until now, actually been able to keep pace with the increase in the global population. In principle, the overall calorie output is sufficient to nourish every currently living person, especially when one takes into account all the food that is not eaten and simply goes to waste. The problems are more closely related to the inequitable distribution and inefficient use of these resources.

Meanwhile, the environmental costs attributable to food production have been all-encompassing, leading to a great loss of biodiversity, excessive use of fresh water, destruction of ecosystems, pollution from pesticides and fertiliser run-off, and emission of greenhouse gases. In addition, the delicate cycles of essential elements such as carbon, phosphorus, and nitrogen have been profoundly disturbed.

These problems were tackled by an international, multidisciplinary group of thirty-seven experts from sixteen different countries under the joint aegis of EAT, a global non-profit start-up dedicated to transforming global food systems, and *The Lancet*, a prestigious family of medical journals. Their task was to answer the question of whether it will be possible to feed a future population of ten billion people with a nutritious diet within planetary boundaries.

According to their findings, agriculture and fisheries are major contributors to changes in the Earth's ecosystems. Agriculture on its own takes up 40 per cent of the planet's landmass, accounts for 30 per cent of greenhouse gas emissions, and uses up 70 per cent of available freshwater. In the case of fisheries, 60 per cent of available fish stocks are already being fully harvested and 30 per cent are being over-fished. The total catch has been declining since 1996. And there are major concerns about expanding aquaculture in its present form on account of its effect on climate, marine ecosystems, and problems of pollution.

Paradoxically, the increase in food production over the past few decades has led to undesirable changes in the human diet in some parts of the world. It has become less healthy because it contains too many calories, is highly processed, and has a growing proportion of food from animal sources. This has led to a rapid increase in diet-related illnesses, such as heart disease, cancer, hypertension, and diabetes. It is thought that over two billion people are overweight or obese. According to the EAT-Lancet Commission report, an unhealthy diet now accounts for more incidents of disease and deaths than those caused by unsafe sex, alcohol, drugs, and tobacco, combined. Yet in other parts of the world millions of people suffer from hunger and starvation.

Change is needed

It is estimated that by 2050 the world population will have increased to almost ten billion. Will it be possible to ensure that there is sufficient nutritious and healthy food for so many people? And unlike our present practices, can it be produced and consumed sustainably? In their report, published in 2019, the members of the EAT-Lancet Commission concluded that the answer to these questions is probably yes, but it will require an enormous science-based coordinated effort to address how food is produced and distributed. They call this *The Great Food Transformation*.

Based on quantitative analyses and careful calculations, the report states that effecting these changes is a formidable challenge, necessitating a plan that is integrated over a number of sectors. It will need to incorporate a reduction of CO_2 emissions from where the food is grown to when it reaches the table, better utilisation of the nutrients in the soil, protection of biodiversity, elimination of monoculture agriculture, regeneration of destroyed arable land, setting aside 50 per cent of the landmass as intact ecosystems, cutting food waste in half, and implementation of regulations governing sustainable use of farmland, water, fertilisers, and other chemicals. These measures would make it possible to establish a sustainable worldwide food system and at the same time meet the development goals espoused by the United Nations.

The authors of the EAT-Lancet Commission report put forward a concrete prescription for a planet-wide healthy diet that would be sustainable and address all of these goals. It is heavily weighted toward vegetables, fruits, whole grains, legumes, nuts, seeds, and unsaturated oils, together with small quantities of fish, poultry, dairy products, and eggs, but very little or no red meat, and only small amounts of highly processed foods, added sugar, and starchy vegetables. Nevertheless, they caution that only a minor increase in the consumption of red meat and dairy products would render it impossible to achieve the environmental goals while still meeting the target of having a sufficiently nutritious and adequate diet to feed the projected global population in 2050.

A model planetary diet

The EAT-Lancet Commission report proposes a so-called planetary menu or diet, recommending that we should each eat 300 grams (10 ½ ounces) of vegetables and 200 grams (7 ounces) of fruit, which is equivalent to five portions, every day. This is supplemented by about 230 grams (8 ounces) of whole grains (rice, wheat, corn), accounting for about 60 per cent of the daily calorie intake, as well as 50 grams (1 ¾ ounces) of starchy vegetables (potatoes, cassava). This is in line with recommendations made by many nutritionists.

The question then arises as to what extent people, primarily in the developed world, who are used to, and enjoy, a diet that is rich in animal products, are really willing, or even able, to eat this much plant-forward food?

Understanding resistance to change

In the efforts to promote plant-based meals it is important to take into account that not everyone is open to new possibilities, never mind making substantial changes to their diet and rethinking their food culture. Some of this can be driven by fussiness—like the proverbial picky eaters who do not like certain things that they have already had, even if only once. Resistance can also be partly due to food neophobia, which is fear that one might possibly not like something that one has never tasted and is unwilling to try. An interesting illustration of this phenomenon is the story, found later in the book, about the reluctance of people in Europe to eat potatoes when they were first introduced from South America in the mid-sixteenth century.

Neophobia is a completely natural mechanism that has conferred an evolutionary advantage by reducing the likelihood of eating something that might be harmful. It can also be due to a person's genetic make-up or to past negative experiences, such as digestive problems or food poisoning, associated with unfamiliar food. Many children go through a phase of neophobia that they normally outgrow by the time they are teenagers, although it lingers at a low level throughout the lives of some adults. Old people may also start to reject certain foods, but that might be due to difficulty in chewing and digesting them or other health-related problems rather than a recurrence of neophobia.

An immediate obstacle to modifying meals to

Fields of kale.

consist primarily of vegetables, fruits, and grains is resistance on the part of children, where neophobia often becomes an issue. Newborns and toddlers are total omnivores and generally eat what they are given or what they see their parents eat. This is a critical stage for two reasons. Some researchers have concluded that the type and variety of food eaten by the mother is reflected in the amniotic fluid and the breast milk and can influence later acceptance of novel food items. Similarly, exposing children to a wide range of solid foods in the very early stages of weaning can have a positive effect. By the age of two, however, children become somewhat wary of food they have not already tried. At this point they have started to move away from their parents to explore the world a bit on their own. This leads to an all-too-common scenario around many family dinner tables when children refuse to eat what is put in front of them. They are told that they are being picky, and the parents try all sorts of strategies to entice them to eat, or at least try, the food, not least because growing bodies need proper nutrition.

It is necessary to accept that children's fussiness about foods and fear of trying unfamiliar ones are not diseases but attitudes that can be overcome. All attempts to do so, however, are not equally successful. For example, experiments have shown that exercising too much control, issuing threats, and taking a hard line often provokes a strong resistance and cements the attitude in place. This is a natural reaction as the children try to assert their independence. A better approach is to involve the children in preparing and serving the food and acting as hosts. Experimenting with new tastes and showing that mealtimes can be enjoyable are also much more productive ways to break down the barriers.

Research has shown that there are small, simple ways to encourage children to eat more vegetables. For example, children between the ages of nine and twelve are more likely to eat raw vegetables that have been cut up into pieces, while those under ten are attracted to vegetables that are served on coloured plates. By way of contrast, repeatedly offering children snack bars with a distinct vegetable component does nothing to increase their vegetable intake. It is in fact not obvious whether being offered a certain type of food over and over leads to an increased willingness to eat it in the long run, even though there might seem to be an effect in the short run.

Children's preferences for a variety of vegetables and, conversely, their rejection of some are highly subjective and pose complex problems for which there are no easy solutions. It seems, however, that allowing children to explore different tastes in an environment geared to them, such as an after-school cooking club, can allow them to take ownership of the question and develop their competencies to make good choices.

One has to bear in mind that children, like many adults, will rarely eat anything simply because they are told that it is 'good for them.' This becomes a real problem if plant-based food in general and vegetables in particular are singled out and associated with tasting 'healthy' rather than being delicious. Nevertheless, this can in some cases be linked to characteristics of the vegetables that can be measured objectively. Chief among these are bitter tastes and unpleasant textures.

Attitudes are evolving

Even though the proposals put forward by the EAT-Lancet Commission are transformative, they can be adopted one step at a time at the level of the individual. There is reason to think that this is already happening and that more and more people, millennials in particular, are embracing plant-forward eating patterns. Although precise data is not readily available, surveys indicate that those switching to vegetarian and vegan diets are doing so both to follow a healthier lifestyle and for ethical reasons, including concern about animal welfare and the environment.

The most recent estimates are that there are about 1.5 billion vegetarians worldwide and about 79 million vegans. On a purely empirical level, this is observable on restaurant menus in Western countries, where many of them now include plant-based options along with their traditional fare. There is also anecdotal evidence in various forms that the trend can have an impact at governmental levels. For example, in 2017, Portugal passed a law mandating all schools, universities, hospitals, and prisons to include at least one vegan selection in their cafeterias. Taking another approach, a number of Dutch municipalities are enacting prohibitions starting in 2024 on advertising meat and dairy products in bus shelters and publicly owned billboards because their consumption makes an outsized contribution to the climate crisis.

Unfortunately, the potential for planetary benefits from the efforts to decrease the consumption of meat and other animal products is partly offset by the availability of highly processed foods that mimic them. Industrial food manufacturers have been quick to identify a potential market. Although these products are promoted as both delicious and having a much smaller carbon footprint than their animal-sourced equivalents, their production is energy-intensive and wasteful and their impact on health is not yet clearly known.

The best, and most scientifically credible, option is to adopt an unprocessed whole food approach. Herein lies the problem. Many people find it difficult to face being served a meal that is predominantly made up of vegetables and grains and to snack only on fruits and nuts. They simply do not think this is really appealing, even though they may 'know' that it is healthy and sustainable. And if they do not like this approach, they will not follow it in the long run.

So, we cannot get around the question of taste and knowing more about how it works if we are to alter deeply ingrained dietary habits.

It is easier to effect changes if one is not dogmatic about eliminating all animal products and instead uses them sparingly. We are convinced that there are ways to meet the challenges of following a sustainable diet that are not beyond what one can easily accomplish in one's own kitchen. They build on acquiring some scientific background knowledge about the nature of the raw ingredients and then combining it with culinary initiative and skilful preparation to turn out delicious plant-forward dishes. The point is that it is not difficult. That is the essence of what follows in the next chapters and in the recipes.

THE BUILDING BLOCKS OF GREEN CUISINE

vegetables and fruits are not enough

Green cuisine is primarily based on ingredients that are not of animal origin. In this book our focus is on the food sources that are popularly thought of as vegetables, according to how they are traditionally used in the kitchen in savoury dishes, but this is where it immediately becomes confusing. Apart from those roots, stalks, and leaves that we always categorise as vegetables, many of the other ingredients, in the strict scientific sense, are not vegetables. They are actually flowers, fruits, nuts, and seeds, while mushrooms and seaweeds are not even plants. Nevertheless, they are all important players, either as the stars of the show or in supporting roles adding umami and other taste nuances.

Of the approximately 435,000 different species of unique land plants, only a few hundred are featured prominently in various kitchen cultures. Rather than using precise scientific terminology, which can become very complicated, we have divided them along culinary lines according to the parts of the plants that we eat. The first group is comprised of those that grow in the soil or along its surface, those that grow above ground, and their flowers. In most cases, the plants themselves are eaten or they die off once the edible parts have been harvested. Those in the second group are either seed-bearing fruits or are nuts or small seeds. They are picked from the stems and branches of plants, bushes, and trees, which are left intact. We then turn our attention to mushrooms and seaweeds, which are not part of the plant kingdom and are classified as fungi and algae, respectively.

What is in a name?

The word 'vegetable' is derived from the Latin *vegetabilis*, meaning able to grow. 'Fruit' also has a Latin root, *fructus*, denoting something that is associated with enjoyment, joy, and satisfaction.

In the ground and along its surface—root vegetables

The edible plant parts that grow right in the ground or along its surface are what we often call 'root vegetables' and these are further subdivided into taproots, tubers, rhizomes, and bulbs. They are particular parts of the root or root system that in the course of the growing season swell up to store nourishment that enables the plants to propagate themselves when they go to seed or set fruits.

A taproot is attached to a single plant and grows upright in the soil, usually with an elongated or round shape. Typical examples are carrots, parsnips, celeriac, and radishes. Tubers are formed at the tips of a root system and can act as clones to grow new plants. Potatoes and yams are well-known edible tubers. Rhizomes, also called rootstocks, are plant stems that grow laterally underground or at the soil surface and can send up shoots to form new plants. Ginger and turmeric are the thickened parts of such rootstocks.

Bulbs are formed when the bottom of the leaf shoots of some plants store up nourishment that is tightly packed into layers in rounded bodies, for example, onions and garlic. In other cases, they form a long cylinder of bundled leaf sheaths, such as leeks and green onions, that grow partly below ground.

Above the ground—greens, leaves, and herbs

Stalks, stems, and leaves are what are often characterised as 'greens.' The stalks and stems are responsible for the transport of water and nutrients to and from the roots and the leaves and provide support for the leaves. A typical example is celery. They are also where flowers and fruits are attached. Leaves are generally more delicate than the stalks and are often quite thin in order to optimise their contact with the air and exposure to sunlight. During the growing season they are packed with chlorophyll, which can give them an intense green colour and enable them to produce energy-storing molecules for the plant. They can also trap a great deal of air in their tissues. Some typical examples are Swiss chard, spinach, and various varieties of lettuce.

Rhubarb is a vegetable

When we eat the stalks of rhubarb plants in dishes such as pies and cobblers, we usually think that we are eating a fruit. But botanically speaking it is a vegetable that grows from rhizomes and is related to the knotweed family.

The term herb refers to the green, soft leaves of plants such as basil, thyme, and oregano that we use primarily to enhance the taste and aroma of dishes. They are generally used in small quantities either fresh or in dried form. It is also possible to forage for wild herbs and weeds such as dandelion greens, wild garlic, and stinging nettles or along the shore for sea asparagus and sea rocket.

Vegetable dishes for days of abstinence

Throughout the Middle Ages and the Renaissance, Catholics were expected to refrain from eating meat on days of abstinence and fasting. On those occasions, their meals might consist only of vegetables, bread, water, herbs, and some salt. As a result, many chefs, especially in France, turned their attention to finding ways to prepare vegetables to preserve their special tastes and turn out appealing dishes. Much later, the famous French chef Marie-Antoine Carême, whose surname coincidentally means Lent, raised the culinary arts to a new level with the skilful use of vegetables, herbs, and sauces. He wrote in a cookbook that it is in meeting the challenges of preparing dishes for Lent that a cook's skill can be seen in a new light.

Swiss chard leaves.

Red onion bulb made up of swollen leaves and a thickened inner stalk.

Bladderwrack with edible fresh tips.

Leaves	Taproots and tubers	Flowers	Edible mushrooms	Fruits	Stalks and stems	Lower stem and bulbs	Rhizome	Seaweeds
lettuce cabbage Brussels sprout leek	potato sweet potato carrot parsnip black salsify	broccoli cauliflower artichoke Romanesco	button porcini Portobello *shiitake* truffle	tomato aubergine chilli pepper bell pepper marrow legumes corn	asparagus celery fennel kohlrabi rhubarb	radish beetroot turnip celeriac onion	ginger turmeric lotus root *wasabi*	bladderwrack dulse *konbu* sea lettuce

Examples of edibles that grow in and above ground and in the sea.

Above the ground—greens, leaves, and herbs

A Catalan celebration of a special onion

The *Calçotada* Festival is held every year on the last Sunday in January in Valls, a small town in Catalonia. It celebrates *calçots*, best described as a cross between leeks and spring onions, which have been cultivated in this area since the beginning of the 1900s. They take on their characteristic elongated shape because they are forced to grow vertically instead of widening into the usual round onion shape. This is done by using a technique called hilling, where soil is mounded up to cover the stalks each time they emerge from the ground. Because they have so little exposure to sunlight, they remain white and sweet. The bulbs are planted from January until March, harvested in the summer, allowed to lie dormant, and then partially cut open and replanted late in the next summer. Two to three small *calçots* grow from each bulb and these are harvested between November and March. *Calçots* have a European Union Protected Geographical Indication, certifying that they are grown according to traditional methods in that particular part of Spain.

Calçots are served by many of the local restaurants during the cold months from November to March. Those that take the dish seriously roast them on an open fire that is fed with large heaps of grapevine clippings and branches. Apart from the onions in a starring role, many offer a standard menu that includes local lamb, sausages, artichokes, citrus fruits, and *crema catalana*, as well as an ice-cold sparkling *Cava* white wine and a local red wine.

An essential condiment is a Catalan tomato-based *romesco* sauce, which is a whole story unto itself. Every chef and home cook has a treasured family recipe and none of them are exactly alike. The sauce is made from the same basic ingredients, but the consistency and spicing vary sufficiently to make it interesting to compare those in one restaurant with those in another. The main difference between the local *calçot romesco* and many other Spanish *romesco* sauces is above all the use of the large, fat almonds, hazelnuts, and small dried peppers, *choriceros*, which are a perfect match for the grilled *calçots*.

Each serving consists of about a dozen or more grilled *calçots*, piled into rounded clay roof tiles. They are eaten with the fingers, a delightfully messy

operation. The outer charred layer is discarded, the white bulb is then swirled in the sauce, slipped into the mouth, and the green end is bitten off.

The festival is a wonderful example of the way in which keeping up a treasured local tradition, centred on a humble vegetable, can help to foster a sense of community. It is a well-organised event run by a group of experienced volunteers. Everyone participates—school children, the adults, the municipal government, and the restaurants. There are contests to judge who has grown the best *calçots* and who can eat the most. On the streets there are places to learn how to make *romesco* sauce and chefs demonstrate their newest creations on a show stage. There is a parade, and the streets are full of life and good feelings, no doubt helped by the wine on sale in the small squares and the large, central plaza that is the focal point of the celebration.

The building blocks of green cuisine

Japanese heirloom vegetables—a specialty of the Kyoto Prefecture

As the number of vegetable and fruit cultivars grown commercially has narrowed down in many parts of the world, it is encouraging to remind ourselves that there are still small-scale farmers who are resisting the trend. An example is a market garden called Higuchi-Noun on the outskirts of Kyoto, which is run by Masataka Higuchi. It has been operated in this location for over 400 years by fourteen generations of the same family. Its specialty is *kyō-yasai*, heirloom vegetables that are particular to the Kyoto Prefecture.

The special care with which these local vegetables are raised is legendary and it goes without saying that they are all raised organically. In order to protect their status, the local authorities created the special designation, *Dentou no Kyō-yasai*, which is analogous to the protected designation of origin labels used in Europe for specialty foods from certain areas. It is a guarantee that the vegetables are cultivars that have been produced locally using traditional methods since at least before the start of the Meiji period in 1868. At present the label covers forty-three different vegetables, among which *mizuna* cabbage, Kuho spring onions, Kamo aubergines, and red Fushimi chilli peppers are especially renowned.

Kyō-yasai are characterised by their wonderful tastes and aromas, unique shapes, and vivid colours. Research undertaken in government laboratories has shown that their mineral, fibre, vitamin, and antioxidant contents are greater than those of other vegetables. Because Kyoto is inland, it was difficult to transport seafood to the area, and in the course of many years, people evolved a local cuisine that focused on vegetables. In particular, they cultivated those that thrive in the special climate of the Kyoto area, which is surrounded by mountains on all sides.

In the traditional calendar used before modern times, Japan's unique seasonal changes were divided into twenty-four sun seasons, each of which is split into three parts of five days each. So, there are really more sun seasons than there are weeks in a year! The sun seasons are still embedded in the food culture, and they are closely followed at Higuchi-Noun. But this means that there is a very narrow window for seeding the various crops. There is also an emphasis on selling

produce exactly when it is in season, at the peak of perfection. In one instance a chef who earlier obtained his produce at the farm discovered that some vegetables, especially turnips (*kokabu*), taste better if they have been worm-eaten. This is due to a number of tasty substances, known as secondary metabolites, that the vegetables produce as a defence mechanism against these predators.

Sadly, it is not a given that *kyō-yasai* will continue to be available in the future. First of all, their cultivation is labour intensive, and this is reflected in their relatively higher price. And some consumers consider the odd appearance of the vegetables strange and off-putting. It is also very difficult to obtain the right seeds for replanting because seed companies are disinterested in producing seeds for these heirloom varieties. The newer types result in a higher yield and, therefore, sell better. As a result, seeds from six to ten different types of plants are collected by hand on the Higuchi-Noun farm. But this is also problematic because it is difficult in such a small area to prevent cross-pollination from other establishments, which can lead to some undesirable hybridisation.

Flowers

Some popular vegetables, for example, cauliflower, broccoli, and broccolini, are eaten primarily for the florets at the ends of the stalks. These are actually collections of tight flower buds that we eat before they have a chance to bloom and go to seed. Bits of yellow on broccoli heads are a sure sign that this is happening. While they are still edible, they are likely to be quite bitter. The large flowers from squash and zucchini are considered a great delicacy. There are also many smaller flowers that are edible and that can be used either as herbs or simply to add a decorative touch. Lavender, borage, calendula, and nasturtium are often found in salads or as garnish on plates.

Fruits, legumes, and pulses

The word fruit is usually associated with something that is sweet and juicy and grows on a tree or bush, such as an apple, a peach, or a berry. In terms of biology, though, a fruit is defined as the body that is formed from the ovary after the plant has flowered and that contains one or many seeds. In some cases, the fruit has a soft, edible flesh in which the seeds are dispersed. Many ingredients used in savoury dishes, such as tomatoes, aubergines, and cucumbers, are filled with seeds and they are actually fruits. In others, such as squash and marrow, the seeds are concentrated in a cavity in the middle of the fruit. Avocados and olives have only a single seed, the pit, embedded, in their soft flesh.

Legumes are a subset of fruits that grow in the shape of an elongated pod that encases the seeds. Some of these, for example, green beans and sugar snap peas, are eaten whole, while others are shelled and only the seeds are eaten. Pulses are also legumes, but the word refers specifically to the dried seeds, such as chickpeas, kidney beans, and lentils. They have a seed coat that can be bitter, and this affects how they are cooked, but they also have excellent keeping qualities. Peanuts are still another example of legumes, having a hard inedible shell that grows underground and usually contains two seeds inside.

Nuts and small seeds

Nuts are also the bodies formed after the plant has flowered and are another subset of fruits that are characterised by having a single seed encased in an inedible shell. Some have a dry, hard shell, such as hazelnuts and acorns. Others have a fleshy exterior that contains a somewhat soft shell covering a seed. Fresh almonds, with a fuzzy pale green flesh that covers the seed in its skin-like shell, are a well-known example. Nuts are sold both in their shells or already shelled and may be eaten as is or roasted in various ways. Seeds, which are also fruits, are mature, fertilised ovules of flowers, generally quite small. They grow and are collected in a variety of ways. Pumpkin seeds are encased in its hollow middle and scooped out from it; sunflower seeds grow on and replace the flower at the end of the stalk; pepper berries are found along the twigs and branches of the plant; and sesame seeds are encased in a pod.

Perhaps it is not widely known that acorns are quite edible. First their hard outer shell must be cracked and discarded. They are then treated to remove some of their very bitter tannins either by soaking them in water or by drying them in warm air. This results in very sweet, aromatic acorn kernels that have retained a great proportion of their antioxidants. They contain about 12 per cent unsaturated oleic acid and of all fruits have one of the lowest glycaemic indices. When the dried acorns are soaked in water, they take on a taste and texture that is reminiscent of almonds. They can be roasted, much like chestnuts, or ground to make flour.

Edible mushrooms and other fungi

The biological kingdom Fungi is home to a vast number of species, many, many more than in the plant kingdom. It encompasses both large complex organisms and unicellular ones.

The large fungi are made up of an underground network of filaments (a mycelium) that can push fruiting bodies containing spores up above the surface. These larger, fleshy fruiting bodies are what we call mushrooms and many of them are edible, although relatively few of them are sufficiently tasty to make a regular appearance on the menu. Some well-known ones are button and oyster mushrooms, Portobellos, and *shiitake*.

By far the majority of the species of fungi are microscopic and unicellular. Many of them are classified as yeasts of one kind or another and some of them have culinary applications. These contain enzymes that can break down proteins, fats, and carbohydrates in other ingredients so that they take on new taste nuances. One particular species of yeast, *Saccharomyces cerevisiae*, is used for brewing and fermenting to convert carbohydrates to alcohol. It is also important in baking and can elicit other taste substances.

Oyster mushrooms.

Dead yeast cells are packed with nutrients and vitamins, as well as taste substances, especially umami. Fungi, together with some algae, are probably the best non-animal sources for making vegetables delicious by bringing out both umami and sensations of *koku*, which will be explained in the next chapter.

Greens from the deep blue sea: microalgae and seaweeds (macroalgae)

The word alga is used to denote a large, heterogeneous group of organisms that come in many sizes and are found in aquatic environments in all parts of the world. These can be roughly divided into two groups. The large ones, the macroalgae, are multicellular and include the more than 12,000 species of marine algae, commonly known as seaweeds, found in all the oceans of the world. The smallest, the microalgae, are unicellular and make up what we call plant (or phyto) plankton. These are found in both fresh and salt water.

Dry microalgae (spirulina) in the form of powder and small sticks.

Very green and nutritious, but forget about taste and texture

A number of the microalgae are edible, and we find a place for them in green cuisine. Some of these are referred to as blue-green algae because of their deep green or blue-green colours, but some of them are actually bacteria (cyanobacteria). Two microalgae, spirulina and *Chlorella*, are readily available where health food products are sold.

Spirulina is the common commercial designation for two species of edible blue-green microalgae, *Arthrospira platensis* and *Arthrospira maxima*, which are bacteria. As their cell walls consist of carbohydrates that humans cannot digest, it is necessary to break them down in order to make the useful nutrients more accessible. This can be done by drying the microalgae in the sun, in a rotating drum, or by freeze-drying, all of which destroy the cell walls. Fortunately, the cell walls of spirulina are easier to break down than those of other microalgae and, to a large extent, this occurs as part of the drying process.

Dried spirulina is typically composed of at least 60 per cent protein, 7–8 per cent fats, and about 15 per cent carbohydrates. Almost 60 per cent of the fats are unsaturated and about half of those are polyunsaturated. The content of super-unsaturated omega-3 fatty acid, DHA (docosahexaenoic acid) and EPA (eicosapentaenoic acid), is almost double that of omega-6 fatty acids, especially gamma-linolenic acid (GLA) and AA (arachidonic acid). It is noteworthy that these microalgae are particularly rich in DHA, actually with about twice as much DHA as EPA. In this respect they are distinctly different from the macroalgae, which are characterised by a preponderance of EPA and virtually no DHA. The super-unsaturated fats, and especially the balance between omega-3 and omega-6, make an important contribution to cardiovascular health and to the proper functioning of the nervous system and the brain.

Spirulina has a high potassium content, almost as great as that of sodium. It is, therefore, a good source of salt for those who are at risk of high blood pressure. It also has an abundance of chlorophyll and carotenoids, which are important antioxidants. In relation to its weight, spirulina is a tremendous source of food energy and nutritional elements. There are about 1500 kJ (360 calories) per 100 grams of dry weight and no food has as great a concentration of proteins.

Because it is so nutritious, spirulina has recently gained a following as a superfood supplement and is also being promoted as a potential change-agent in the fight against malnutrition in poorer countries. But this is really just a rediscovery. For centuries the Aztecs harvested blue-green algae in the enormous lake, Texcoco, that surrounded Tenochtitlán, the capital of their empire. It was dried in the sun to produce cakes that could be mixed with food, baked, or eaten as is. Sadly, the Spanish conquistadors drained the lake and the consumption of spirulina in Mexico disappeared from the historical record. In Africa, however, women living around the shores of Lake Chad continue to harvest spirulina as they have done for centuries, using simple climate-friendly methods. This not only helps them to feed their families, but also provides them with a much-needed source of income as they can sell the finished cakes to health food producers for a good price.

Another genus of edible microalgae is made up of several species of *Chlorella*, which are plant-like, single-celled organisms that carry out photosynthesis and, as a consequence, have an intense green colour. They are almost as protein-rich as spirulina and also contain significant quantities of super-unsaturated omega-3 fatty acids.

Both spirulina and *Chlorella* are sold in the form of powder or small green pills. They taste somewhat dry and a little like seaweed, but with a slightly sulfuric edge. Neither of these microalga types contributes interesting tastes, aromas, or textures to a dish. Apart from their nutritional value, their main gastronomic contribution is to add intense colour, for example, when they are blended into a smoothie.

Seaweeds—an integral part of the food web

The term seaweeds is used to refer to a diverse group of green, brown, and red macroalgae that fall into distinct taxonomic categories. Because of their name, however, many people think of them as belonging to the plant kingdom (Plantae) or refer to them as 'sea vegetables,' even though they are only remotely related to plants.

While seaweeds often have a superficial resemblance to plants, the tissue of the majority of them is built up very differently from that of the higher forms of plant life. They do not have leaves and stems in the botanical sense of the words, nor do they bloom, produce seeds, or set fruit. Seaweeds do not need to have roots as their cells are in direct contact with the surrounding water from which they derive their nourishment. Many do, however, fasten on to cliff faces or stones, while others float about freely. A few species have evolved a system for the internal transport of vital salts and the products of photosynthesis, but the majority of them are basically a collection of undifferentiated cells that are responsible for looking after their own needs.

The macroalgae are to the sea what forests, underbrush, bushes, and groundcover are on land. But they are also an important food source for many marine animals and are harvested widely for human consumption in many parts of the world. Although they are located at the bottom of the aquatic food web, for the above reasons alone it is important that they are factored into worldwide deliberations about how to create sustainable food systems.

Seaweeds are large marine macroalgae that grow in all the oceans of the world.

Shōjin ryōri—*taking green ingredients to a higher level*

Shōjin ryōri is a special vegetarian culinary tradition that has evolved in the monasteries of Japanese temples. Like many Asian dietary practices that are strongly linked to religious traditions—including those of Hinduism, Jainism, Sikhism, and Buddhism—it prohibits, either partially or totally, the consumption of animal products, especially meat. It is probably the most highly developed of these cuisines in that the monks have found several different ways to prepare dishes that are both vegan and also satisfy our natural, biological desire to eat food that is rich in umami, all without using any animal products and with a focus on green and seasonal local products.

One of the central precepts of Buddhism is that one must not kill any sentient being. But it is important to understand that this is a recommendation rather than an absolute commandment. Believers are expected to use their discretion to follow it in an overall context of what is possible in practice and to prepare meals in a way that shows utmost respect for the natural world. For example, those living in a very cold climate such as the mountains of Tibet may, to a certain extent, eat meat. There is also an emphasis on striving to use a raw ingredient as fully as possible so that nothing potentially edible is thrown away. This way of thinking pre-dates by hundreds of years the current focus on ecology, sustainability, and reduction of food waste.

The prohibition against eating meat undoubtedly gave impetus to finding ingredients that could contribute umami to make the food more palatable. Dried and fermented fish, seaweeds, pickled vegetables, soybeans, and fungi were a partial solution. Another was the use of *dashi*, the ubiquitous broth made with kelp and dried bonito flakes, that adds an abundance of complex umami tastes to vegetable dishes. Within the Zen school of Buddhism the consumption of fish is banned and it is not permitted to make *dashi* with the fish flakes, *katsuobushi*. They were therefore replaced by dried *shiitake* mushrooms, which interact synergistically with the *konbu*, resulting in *shōjin dashi*. This was how the distinctive, strictly vegetarian *shōjin ryōri* began to take shape.

The literal meaning of the expression *shōjin ryōri* points to the depth of its connection to the plant-based cuisine. *Ryōri* means cuisine or preparing food. The Chinese symbol *shō* denotes a spiritual outlook, endeavour, striving, or being energetic in a certain direction (focus). Similarly, *jin* means to advance forward. The combination of these two symbols can be interpreted as zealously devoting oneself to something in a spiritual way by giving up something, in this case animal products. As the followers of Zen Buddhism are hoping to achieve a higher state of consciousness, *shōjin ryōri* is often simply described as 'the enlightened kitchen.'

This temple cuisine is basically vegetarian and, in many cases, vegan as well. From a point of view focused on taste, the core of Zen Buddhist temple cuisine can be seen as a striving after good taste and umami, all without meat or fish. Not all practitioners of *shōjin ryōri* follow a strictly vegetarian diet, but also include fish and other non-vegetarian elements in the food, not least to make a classic *dashi*.

It is also interesting that some who follow the precepts of *shōjin ryōri* will openly admit that they often have a need and desire to eat non-vegetarian food, especially meat and fish, although they resist the temptation. One response to this is a style of cooking called *modoki ryōri,* which features dishes made with plant-based ingredients that mimic the taste, look, and texture of their traditional non-vegetarian counterparts, for example, sushi, sashimi, and grilled eel. It can be traced back to a Chinese sect of Zen Buddhism (Obaku) that introduced a variant of temple cuisine called *fucha ryōri* to Japan in the mid-seventeenth century.

Shōjin ryōri chefs are trained to draw out umami from exclusively vegan ingredients—vegetables, seaweeds, cereals, legumes, wild plants, and mushrooms—with the help of *dashi*, made with either *katsuobushi* or *shiitake*. Creating delicious tastes in this way presupposes that the chef has an intimate knowledge of the raw materials and how they impart umami when handled with both understanding and expertise. An example is lotus root, which is rich in umami. Other important ingredients are soy sauce (*shōyu*), *miso*, and *fu* (wheat protein).

The secret at the heart of *shōjin ryōri* is an approach to cooking that is essentially the opposite of that of Western cuisine. In much of the latter, the savoury aspect of a dish depends on adding spices and herbs in the course of preparation or at the very end, just before it is served. Many recipes sign off the instruction, "Season to taste." In temple cuisine, the taste elements are introduced right at the beginning and normally lead off with *dashi*, the soup stock which is almost pure umami. It works its magic on vegetables and prepared products such as tofu, *yuba* (the skin from soybean milk), *fu* (wheat gluten), cooked rice, and soups. Sometimes it is even incorporated into desserts.

The perfect gustatory balance of *shōjin dashi* is achieved by the addition of dried *daikon*, a little salt, soy sauce, *mirin* (sweet rice wine), *yuzu* (Japanese citrus fruit), toasted tofu, and *shichimi*, a special seven spice combination. Even though there are local variations, *shichimi* is traditionally a mixture of *sansho* pepper, white and black sesame seeds, red chilli pepper, dried ginger, *ao-nori* (a green alga similar to sea lettuce), dried citrus peel, and hemp seeds.

Another aspect of *shōjin ryōri* is the influence of the changing seasons.

Apart from the traditional four seasons, the year is divided into seventy-two micro-seasons, many of them characterised by when a given raw ingredient is at its peak. It is not unusual to eat a dish with a particular vegetable or herb at a Japanese restaurant only to be told when trying to order it the following week that it is off the menu because it is no longer in season.

Shōjin kaiseki is a very high-level version of ordinary temple food. In contrast to the much simpler daily monastery fare, which includes only three dishes—a little soup, a bowl of cooked rice, and an accompanying side dish—*kaiseki* consists of many. The Japanese expression for this elaborate meal embodies elements of respect (*kai*) and cleanliness (*sei*), an indication that it was never intended for the monks, but for their important guests, often from aristocratic circles. In *cha-kaiseki* it is combined with the traditional tea ceremony.

Within the strict guidelines that govern *shōjin ryōri*, the number five takes on a particular and somewhat mystical role. Efforts are made to achieve balance and harmony in each meal by incorporating five colours (green, white, red, black or dark, and yellow), five tastes (sour, sweet, salty, bitter, and umami), five methods of preparation (raw, steamed, boiled, simmered, and roasted), and five elements (water, wood, fire, earth, and metal). Umami is a more recent expression replacing the earlier concept of spicy or strong taste. In some contexts, there is also mention of a sixth taste, which is characterised as mild or delicate. Actually, classical *shōjin ryōri* dishes generally have a mild taste because they are not seasoned with strong spices or pungent ingredients such as onions, garlic, and leeks. This is in keeping with the focus on purity, simplicity, and respect for the raw ingredients by bringing out their inherent flavours.

Because it is believed to align the mind, body, and soul, there are five rules related to Zen mindfulness that should be followed when eating a *shōjin ryōri* meal:

1. Appreciate the effort that has gone into putting the food on the table and acknowledge those who have prepared the meal.
2. Think about your own deeds and whether you are worthy to partake of the meal.
3. Do not eat too much and be humble and satisfied with what you are offered.
4. Think of the food as a medicine that will determine whether you can maintain good physical and mental health.
5. Eat the meal as part of a way to cleanse yourself.

Even though *shōjin ryōri* is prepared according to very strict practices and steeped in traditions that are centuries old, it is still served in its classical form in some temples as well as in restaurants. But modern chefs are seeking renewal and finding ways to innovate. Their principal aim is to introduce variety and create new taste impressions, experimenting with a broader range of vegetables. This development is especially important for Japanese restaurants in other parts of the world that may need to adapt aspects of this venerable cuisine in order to use fresh, locally sourced ingredients in season.

Bananas and their rhizomes—a cautionary tale

Bananas are among the world's most widely consumed fruits, an estimated 100 billion bananas a year. There are more than 1,000 varieties but over half of them are consumed locally. At present only one cultivar is commercialised and grown on a large scale for export, primarily to Europe, the United States, and Asia. This is the Cavendish banana, which descends from plants originally grown in Mauritius that had been cultivated in the greenhouses of William Cavendish, the 6th Duke of Devonshire. This cultivar now accounts for about 47 per cent of the total production and is an export source of income for Ecuador, Philippines, Costa Rica, Colombia, and Guatemala. Sadly, the Cavendish banana may now also be on the verge of extinction. The story of how this came to be can held up as an almost textbook example of the pitfalls associated with lack of diversity and monoculture. Wild bananas reproduce both sexually and asexually. In the case of the former, after their flowers are pollinated, they swell up to set bananas that have seeds. These fruits often have very little flesh, and their seeds may be large and hard. But because this is sexual reproduction, mutations from cross-pollination may occur and undesirable hybrids can be created. Asexual reproduction happens after the plants have fruited, when they send up suckers from the rhizomes that are part of their root system. These in turn grow to replace the original plant, which then dies off. The popular eating varieties descend from mutant plants in which the seeds appear only as vestigial tiny, dark dots that cannot germinate. They can only be propagated by cultivating the shoots sent up by the rhizomes and allowing them to grow to maturity. This is efficient, but as all the individual plants are clones, genetic diversity is non-existent, and they are equally vulnerable to disease. A pathogen that can kill one plant can kill them all.

Until the first half of the twentieth century, the dominant banana cultivar was called Gros Michel, which was much tastier than the Cavendish strain, had a thicker skin, and did not bruise as easily. Then disaster struck in the form of a fungus, *Fusarium oxysporum*, which is thought to have originated in Southeast Asia in the late 1800s. It came to be known as 'Panama disease' and eventually spread to all banana growing areas. The fungus produces spores that enter the root system, travel through the xylem, and eventually cause the banana plant to turn yellow, wilt, and die. But what happens below the ground does not stay below the ground. The fungus can lie dormant in the soil for up to forty years, is found in the water and soil, gets on workers' boots and farm vehicles, is resistant to fungicides, and is virtually impossible to eradicate. By the time it swept through huge plantations in Central America in the 1950s it had already caused enormous losses and virtually eradicated those banana crops.

In desperation and in need of a quick solution, growers replaced the Gros Michel variety with the Cavendish, which was resistant to Panama disease, even if not as tasty and less sturdy. It had qualities that were ideal for the export trade—it was

well-suited for transport by ship because it stayed green for weeks after harvest, gave an abundant yield, grew to a uniform size, and looked attractive when ripe. The supply chain, including how they were boxed, was tailored to take advantage of these specific traits of the Cavendish.

All seemed to be well until 1990 when an outbreak of another variant of *Fusarium oxysporum*, to which Cavendish bananas are not resistant, was identified in Taiwan. It has now been found intermittently in many other locations, including Australia, Africa, and South America. This is a particularly serious threat to food security in Africa, where bananas and plantains, the bananas that are used mostly for cooking, which are also susceptible to Panama disease, are a staple part of the diet of millions of people. Intensive efforts and funding are being poured into finding microorganisms that could combat the fungus, as well as portable testing kits to detect it before it has a chance to spread further.

The current market standards and export infrastructure are not set up to handle multiple banana cultivars and it would be expensive to retool the whole operation. Furthermore, thousands of people working in the large plantations depend on them for their livelihood. Consequently, money is instead being poured into research focusing on developing a disease-resistant variety that closely resembles the Cavendish using traditional breeding methods, but this could take many years. Meanwhile, in Australia researchers at the Queensland University of Technology have fast-tracked the development of a genetically modified Cavendish banana that does not succumb to the variant of Panama disease that is currently in circulation. Their next goal is to use gene-editing technology, CRISPR, to develop a version that is not genetically modified, thereby possibly avoiding consumer opposition to GMOs.

To base half of the harvest of a major food crop on cuttings from the rhizomes of a single cultivar has proven to be exceptionally risky. It seems reasonable to ask whether turning to plant breeding and biotech options that might, once again, prioritise monoculture practices is the most logical approach.

UNDERSTANDING THE TASTE EXPERIENCE

taste, taste-like sensations, and aroma, too

As we have already pointed out, the biology of plants works at cross-purposes to human taste preferences, which are driven by evolution, to make the prospect of a plant-forward diet seem less palatable. Fortunately, there are many effective and relatively simple ways to address these challenges so that the taste of green ingredients becomes appealing to most people, even those who say that they simply do not like vegetables.

In order to understand the underlying principle of how we might overcome the reluctance to adopt a diet that incorporates more green cuisine, it is important to note that this involves the whole 'taste experience' not just 'taste' *per se*. Identifying the factors that determine how we react to what we eat is the key to preparing delicious food.

Our impression of how an ingredient or a dish 'tastes' goes far beyond what our taste buds tell us. It is just as dependent on other sensory perceptions—how the food looks, smells, and feels in the mouth, and whether there is any sound when we bite into it or chew it.

The taste experience can be further analysed in two ways. One aspect is the taste of the ingredient itself, either raw or lightly prepared, for example, by steaming. The other is how it compares with the tastes and textures that are added or created by more extensive treatment such as boiling, drying, fermenting, or marinating.

There are significant differences in the taste and smell of various plants and of the parts of which they are composed. The same is true of mushrooms and seaweeds. Nevertheless, there are some common characteristics for each of these groups that depend on the biology of the species and the natural environment in which they grow. We will describe these aspects in greater detail after we have had a closer look at what is meant by the 'taste experience.'

The basics of taste

What we commonly call taste is usually based on a combination of several sensory impressions. These all come together in the brain where they are interpreted and integrated into what we could refer to as a taste experience, which is multisensory. When we eat, all five senses—taste, smell, touch, sight, and hearing—are normally involved. Our overall impression is very rarely driven by a single sense, especially when we eat any of the green ingredients. But in these cases, it is often our reaction to the texture that is dominant.

Many plants leave us with sensory impressions that we interpret as tastes, but that actually have nothing to do with input from our taste buds. We might call them 'taste-like sensations.' They are instead driven by other chemical or physical reactions. For example, we say that chilli pepper, black pepper, horseradish, and mustard have a taste that is strong, burning, or irritating. These are not true tastes but express what is known as a trigeminal sensation (chemesthesis) that occurs when certain substances in the raw ingredients irritate or damage certain nerve endings in the mouth. This type of taste impression is actually an aspect of mouthfeel.

We also say that some fruits, such as acidy grapes, unripe persimmons, and still green bananas have an astringent taste. This is, likewise, not a real taste, but a particular reaction on the part of the mucous membranes in the mouth.

Our individual evaluation of the taste experience (the physiological impression) is linked to, but is not identical to, whether or not we like the food (the hedonic response) and possibly whether we have a preference for that particular dish. Here a whole host of other, more aesthetic and socio-psychological factors, such as tradition, culture, and our anticipation of how the food will taste or memories of having eaten it, come into play.

The hedonic response is influenced by social interactions to such an extent that a certain serving of vegetables eaten in different social settings can be perceived as both something that one likes or does not like, even though the taste itself is identical.

Horseradish is associated with strong, burning taste-like sensations.

True tastes and 'taste-like sensations'

The five basic tastes

A true (physiological) taste is perceived by the taste buds on the tongue, which have specialised receptors that can identify each of the five basic tastes: sour, sweet, salt, bitter, and umami. By definition, all true taste impressions are a function of these five, either on their own or in combination. And, by extension, it follows that a basic taste cannot result from a combination of the other four, for example, umami cannot be elicited by putting together sour+salt+sweet+bitter.

Mouthfeel and texture

Mouthfeel is a part of what is called the somatosensory system, which is stimulated by the perception of physical effects (tactile stimulation) such as pressure, touch, movement, vibrations, and temperature, and in some cases pain. Lips, tongue, teeth, and the entire oral cavity are all involved in assessing the structural properties of a particular food, what we describe as its texture. It is considered to be a sensory property of the food and, consequently, one that can be recognised and described only by a human, as opposed to a physical property that can be measured in a laboratory.

Normally, texture is perceived by several senses and, therefore, cannot be characterised unambiguously by a single descriptor such as hard or creamy. A typical example is food that is considered to be crunchy. Crunchiness encompasses both the feeling in the oral cavity and between the teeth of a material that breaks apart when we chew it and the quick vibrations and associated sound that is picked up by our hearing, possibly even through the bones in our skull.

Chemesthesis: burning, hot, and irritating

Chemesthesis is a concept used to describe the sensitivity of the skin and mucous membranes to chemically induced reactions that cause irritation or pain and that may damage cells and tissues. In the mouth this is registered as a sharp taste when we eat foods such as chilli peppers, which contain capsaicin, black peppercorns, which contain piperine, or horseradish and mustard, which can release isothiocyanate. As the endings of the trigeminal nerves (the fifth paired cranial nerves) are the ones affected, chemesthesis is sometimes referred to as the trigeminal sense. It can be thought of as a warning signal that there is something in the food that may be harmful to us.

Sensations of temperature are also related to chemesthesis, in that certain chemical substances can interact with and, in a way deceive, some of the temperature- and pain-sensitive nerves associated with mouthfeel. This gives rise to false perceptions of heat and cold that are not directly related to the actual temperature of the food. Capsaicin induces a burning sensation, whereas menthol, peppermint, and camphor all feel cold in the mouth even though their temperature has not changed.

Astringency

At one time astringency was considered to be a basic taste. But we now characterise it as a mechanical sensation, very similar to mouthfeel, caused by substances with a high concentration of tannins (polyphenols). A few well-known examples of astringency are the way the mouth puckers up and feels dry as if sticking together when we drink very young red wine or strong tea or eat unripe fruit.

The sensation of astringency is due to chemical reactions between the food and the surface of the tongue and saliva. Proline-rich proteins in the saliva bind with the tannins from the food, causing them to clump together. These aggregated proteins feel like tiny particles and make the saliva more viscous, reducing its capacity to help the food slide over the tongue and past the sides of the oral cavity. The parts of plants that are bright red, violet, or blue contain many polyphenols in the form of anthocyanins, which is why vegetables with these pigments often have an astringent effect.

The taste of plants

To a certain extent, our taste impression of plants is driven by the basic tastes, but we are also influenced by the aromatic substances that we associate with them, especially fruits. Nevertheless, texture is almost as dominant and other important taste-like sensations, especially irritation and astringency, also play a part.

Taste and aromatic substances are found mostly inside the cells of plants, and they are released to the greatest extent when the cells are broken open, for example, by slicing and chopping the vegetables. This initiates a cascade of reactions when the cells are destroyed, and the enzymes are set free. In turn, this leads to a whole range of new taste and aromatic substances, of which some, such as sweet ones, are desirable and others are not. Cooking can further modify the aromatic substances and give rise to the well-known taste and smell of overcooked vegetables.

Life after death
The taste, aroma, and texture of fresh fruits and vegetables change after they have been harvested, which can be a positive thing, for example, when certain fruits continue to ripen. In by far the majority of cases, however, these characteristics are altered in ways that are less desirable, for example, when a bruised apple becomes dry and mealy. The outcome is very dependent of the type of plants in question and how they have been stored. Temperature, light and moisture, preservation, and the time of harvest are all key factors.

Even though they have been harvested, plants and their constituent parts are still alive. As long as their cells and enzymes have not been damaged, for instance by heating or preserving, and as long as microorganisms in the environment in the form of bacteria and fungi have not completely broken down their cells, some of their vital functions carry on.

This leads to the age-old question of whether or not a fresh carrot that one has just pulled out of the ground and is in the process of eating is still alive. The distinction between the living and the dead state is much less clear in the case of plants than in foods sourced from animals. Potatoes and onions that have been harvested in a given season can even be considered to be in a form of hibernation from which they emerge if replanted in the next season.

The metabolic processes of the harvested, but still live, plants continue for a certain period of time. But as the plants no longer have an ongoing source of nutrients, they gradually deplete their energy depots and accumulate waste products, both of which affect their taste and aroma. At the same time, their texture can undergo changes, such as when asparagus and broccoli become woody. Fruits and vegetables will also inevitably lose water content, causing their turgor to decrease and leaving some of them flaccid and soft.

It is possible to counteract these developments to a certain extent by storing the plant matter at a low temperature so that their own enzymes work more slowly. Another way is to control the environment with regard to moisture, oxygen, carbon dioxide, and nitrogen. A simple kitchen technique, blanching, can stop the enzymes in vegetables from working and preserve their taste, aroma, and texture for a short period of time.

Taste
Of the five basic tastes, it is especially sweet, sour, and bitter that are more pronounced in fruits and vegetables. Less commonly, certain plants that grow in marshy, saline environments, such as sea asparagus (*Salicornia europaea*), are very salty. And, as noted below, some vegetables and fruits are also rich in umami.

Most of the sugars found in plants are stored in the form of starch, which is tasteless. Unripe fruits are typically very sour on account of their large content of organic acids such as citric, oxalic, malic, and tartaric acids. It is only after fruits have ripened that the starch is converted to sugar, drawing out the sweet tastes. Vegetables can also be slightly sweet, but the taste disappears shortly after they have been harvested, as does the somewhat sour taste found in the green leaves and stems. These are otherwise characterised primarily by their bitter tastes, although selective plant breeding has resulted in modern species where they are less prominent. In the case of vegetables, therefore, the best and most varied taste impressions are to be had from eating them while they are still very fresh.

Umami as a separate taste in vegetables and fruits deserves a special mention because, as already noted, its absence is one of the challenges associated with green cuisine. Nevertheless, free glutamate, which is the source of umami, is found in varying quantities in tomatoes, strawberries, oranges, walnuts, potatoes, corn, and green peas, among others.

When it comes to sources of umami in the plant world, however, tomatoes are in a category by themselves. This is particularly true of dried, sun-ripened tomatoes, which are very rich in umami. The glutamate content of a tomato increases dramatically as it ripens in the sun, having ten to fifteen times more when it is red than when it was green. Another special characteristic of tomatoes is their considerable content of free nucleotides, especially adenylate, which has a strong umami synergy with glutamate. In addition, there is about three times as much glutamate and adenylate in the inner seedy part of a tomato as in the outer fleshy part. The way that tomatoes can contribute umami and create umami synergy is probably the most important reason why it is such a central vegetable in many different food cultures.

Koku and *kokumi*

Koku is a hard-to-define Japanese expression for a special concept, associated with a taste attribute that combines elements of continuity, mouthfulness, and complexity. It can enhance the sensation of umami, sweet, and salty and at the same time suppress bitterness. *Koku* is usually associated with foods sourced from animals, especially scallops, fish sauce, certain cheeses, and fermented shrimp paste. But it is also ascribed to fermented beans in the form of soy sauce and can be elicited when garlic extracts interact with other foodstuffs. Taste aspects related to *koku* are denoted by the term *kokumi* and those substances (specific peptides) that elicit *koku* are called *kokumi* substances. One of the substances found in these examples is glutathione, which has also recently been identified as present in large quantities in some fungi, especially porcini mushrooms (*Boletus edulis*).

What is umami and synergistic umami?

Although umami is a basic taste in its own right that cannot be created by the other four tastes in combination, it is able to have an effect on them. It turns out that umami can amplify perceptions of saltiness and sweetness and can also suppress bitterness. Umami is most pronounced when two other types of substances are found in the food.

What is known as basal umami is elicited by free glutamate, a sodium salt of the amino acid, glutamic acid, which is found in proteins in the food ingredients. One could say that glutamate is to umami what table salt (sodium chloride) is to saltiness.

Free nucleotides (5'-ribonucleotides), which are derived from the breakdown of nucleic acids, are also important. These nucleotides, especially inosinate (IMP), guanylate (GMP), and adenylate (AMP), interact synergistically with glutamate to enhance umami to an even greater extent. Glutamate, and consequently basal umami, is prevalent in seaweeds, fish sauce, aged cheeses, sun-ripened tomatoes, air-dried ham, potatoes, and green peas. Synergistic umami is abundant in fish, shellfish, and meat in the form of inosinate, in dried mushrooms, for example, *shiitake*, in the form of guanylate, and in the form of adenylate in scallops and tomatoes.

Just a small quantity of a substance that brings out synergistic umami can almost magically increase the taste of basal umami. This effect is in play when we season a sauce with a little blue cheese or tomato paste, combine meat and vegetables to make a soup stock, or add ripe tomatoes, anchovies, and Parmesan cheese to a salad. The magic of umami synergy is the most powerful tool to make vegetables delectable.

Garlic as a 'fixer'

Garlic has a wonderful way of interacting with other ingredients to enhance their taste, and in the case of vegetable dishes it opens up some very particular possibilities to add both umami and *koku*. The characteristic taste and smell of garlic is due to sulfur-containing compounds that are formed enzymatically in great quantities when the garlic is cut or damaged. A substantial proportion of the taste and smell impressions are dissipated when the garlic is heated over a long period of time.

Garlic contains one of the *kokumi* substances, glutathione, that is known to be responsible for that particular sensation of *koku* that is described as having both continuity and a great deal of mouthfulness and is more delicious. This may well be why garlic has been used for millennia and is so prevalent in many different food cultures. It is a veritable 'fixer' for those vegetable dishes that have little umami and ability to elicit *koku* on their own.

Black garlic, which is sometimes incorrectly called fermented garlic, is produced by storing garlic bulbs at 60–80°C (140–175°F) in a closed container with controlled humidity level of 70–80 per cent for several weeks. The cloves turn pitch black, and their rather pungent taste becomes mild, a little acidic, sweet, and pleasant, rounded out with flavour notes of balsamic vinegar and tamarind. This process is sometimes referred to as fermentation, but the changes are actually due to very slow-acting, low-temperature Maillard reactions that result in the distinctive dark and black browning by-products. At the same time, the texture of the garlic takes on a soft, creamy, and slightly waxy consistency that serves to accentuate its characteristic *koku*.

Koku attributes and *kokumi*

While studying the taste components in garlic extracts in 1990, a Japanese researcher, Yoichi Ueda, and some of his colleagues made a breakthrough discovery. Although the extracts themselves had very little taste, they were able, even in small quantities, to enhance other tastes and in that way heighten the overall taste profile of a food. The Japanese call this particular property a 'taste attribute,' in this case *koku*, denoting that the taste experience is characterised by mouthfulness, richness, and continuity. The resulting taste impression was one of deliciousness and harmony that could not be produced by any of the five basic tastes. As the researchers had found that the garlic extract was a source of *koku* in an umami solution, Ueda suggested that this special taste phenomenon was induced by special so-called *kokumi* substances.

Even though it describes a distinct, describable effect, it is rather difficult to define and understand the concept of *koku* and there is possibly some overlap between the *koku* attributes and umami. The expression encompasses a complex interplay of the various taste impressions that seems to make them more concentrated and works in tandem with a physical sensation of filling and coating the whole mouth. It also implies continuity, meaning that the taste experience is long-lasting, becomes more intense over time, and then tapers off slowly. Even though it is not an independent taste, *koku* is associated with making food more delicious.

The question then arises as to the source of *koku*. It is now known that there are small bits of proteins, called dipeptides and tripeptides, that elicit *koku* in a range of foods. These peptides are called *kokumi* peptides. Dipeptides have been found to interact synergistically with umami taste substances. For example, some of the more active ones are found in Gouda and Parmesan cheeses, soybeans, and yeast extracts.

It has also been demonstrated that the sensation of *koku* can be produced when certain tripeptides stimulate the calcium channels on the epithelium of the tongue. Very prominent among them are gamma-Glu-Cys-Gly (glutathione, GSH) and gamma-Glu-Val-Gly. Glutathione is generally present in only small quantities in dairy products, cereals, and bread, in moderate amounts in fruits and vegetables, especially garlic, and in somewhat greater quantities in raw meat, for example, beef, and chicken. But even in miniscule quantities (2–20 parts per million) it is very effective in eliciting *koku*. Gamma-Glu-Val-Gly, which is found in abundance in fish sauce, soy sauce, shrimp paste, scallops, and beer, is probably the most interesting peptide because it is thirteen times more effective than gamma-Glu-Cys-Gly.

The *kokumi* peptides that trigger the sensation of *koku* have no taste of their own, but can suppress bitterness and enhance salty, sweet, and umami tastes. It is still unclear how they interact with sour tastes. Their ability to activate the sensation of *koku* has been studied most closely in connection with sweet dishes and soup stocks made with meat and vegetables. The objective is to reduce the sugar, salt, and fat content of such foods without sacrificing taste.

Much research is now being undertaken both in academia and in industry to find ways to unlock more of the substances that result in an impression of *koku*, especially from foods that are plant-based. There is particular focus on using fermentation and enzymatic treatments to break down proteins found in plants and dairy products to form the desired peptides. This knowledge, coupled with an understanding of the synergy between umami and *koku*, can be put to work to create vegetable dishes that are more palatable, have enticing aromas, and appeal to a broader cross-section of the population. It can also help to meet the growing consumer demand for plant-based foods driven by an awareness of the need to adopt healthier and more sustainable diets that lighten the burden on the Earth's resources. As a result, *koku* is coming into its own as less of a nebulous concept and more of a mainstream idea.

Ajinomoto—where umami and kokumi seasonings were launched

Kawasaki, an industrial city at the southern edge of Tokyo, is home to factories and research and development laboratories belonging to Ajinomoto. This corporation, which is regarded as one of Japan's stellar enterprises, has operations in thirty-five countries, and it is one of the world's biggest food and biotechnology conglomerates. Of special interest in connection with the topic of this book is the production of a range of seasonings and enzymes for the food industry.

Ajinomoto, which can be translated to mean something like the 'innermost essence of taste,' is a reflection of the company's origins. It was founded in 1908 by a Japanese professor of chemistry, Kikunae Ikeda, in conjunction with Saburosuke Suzuki, at the time one of the country's most important industrialists. The teamwork between the research scientist and the enterprising businessman proved to be a stroke of luck. Earlier that year, Professor Ikeda had identified the substance, monosodium glutamate (MSG), which is responsible for the delicious taste of *dashi*, the soup stock that is ubiquitous in Japanese cuisine. He quickly realised that this ingredient had its own taste, which he posited to be a fifth basic taste and called it umami. He also saw that MSG had commercial potential and patented his discovery, which he was able to bring to market through a company that was created as a subsidiary of Suzuki's existing firm.

Umami Science Square, a small museum and exhibition space in Kawasaki, attests to the importance of Ajinomoto. Here, the story of Professor Ikeda's serendipitous discovery of the fifth basic taste and the ensuing industrial growth of the company over the next 110 or so years are outlined in simple, very accessible displays. There is even a flask with the very substance (MSG) that Ikeda isolated from *konbu*, the macroalga that is traditionally used when making *dashi*.

It took almost a century for umami to be accepted as a true basic taste, but during the whole time the researchers at Ajinomoto were able to investigate all possible aspects of umami. And for the past three decades, they have also turned their focus to finding out more about *koku* and *kokumi* substances. They are still grappling with the problem of how to come up with a clear definition of the concept. To that end, Japanese researchers working in the field of taste and

The new established Ajinomoto factories in Tokyo in 1914.

technology stage annual symposia where they review progress in this area and discuss the possibilities of achieving a science-based consensus. A major problem seems to be terminology and how to formulate a definition that is distinct, even though it cannot avoid overlapping with other definitions, such as the one for umami.

Given that coming up with precise descriptions related to *koku* poses a challenge even for the Japanese, it is easy to understand how difficult it is for others who do not have concepts of either *umai* (in umami) or *koku* (in *kokumi*) embedded in their everyday vocabulary to grasp the idea. Nevertheless, finding a way to do so is of great importance in order to be able to train the professional tasters and taste panels who will evaluate whether a given food product can create an impression of *koku*. Another outstanding question is how *koku* interacts with mouthfeel and the texture of a particular food. Some of the ingredients being studied specifically with respect to *koku* are soups, aged cheeses, sausages, green tea, and a variety of fermented marine products.

Aroma

The aromatic substances in plants are found mostly in the vacuoles of their cells. There is an enormous variety of types and intensity of these substances, ranging from the mild smell of peas and potatoes to the very strong ones from onions and cabbage. The smell of some fruits can be so pronounced and characteristic that we actually conflate it with a taste. For example, we sometimes refer to a 'strawberry taste' even though there really is no such thing.

The aroma associated with a fruit or a vegetable is composed of a number of substances, often a very large number, that are volatile, airborne, and can be inhaled. The characteristic smell of a plant is only rarely due to either a single or a few of these aromatics. What this means is that when two plants have one or more of these substances in common, it can be hard to tell them apart, even though their composite odour profile is distinctly different. For example, the aromatic thymol is found in both oregano and thyme, making it difficult to decide which herb is which based on smell alone. One needs the other for comparison purposes to make a correct identification.

Aroma families in plant foods

Harold McGee, an American food science writer, is best known for his popular book *On Food and Cooking*, which describes the cultural history and science related to raw ingredients, food, cooking, and gastronomy. In this book he discusses the aromatics in plants, dividing them into a series of aroma families, naming them after the characteristic smells that we often associate with a fruit or a vegetable. In the majority of cases, these compounds are formed by the action of enzymes either in the fresh fruit or vegetable or when the plant is damaged in some way, for example, by cutting it open. McGee has expanded on this work in his recent book *Nose Dive*, which delves deeply into the world of smells from every conceivable angle.

Aroma and evolution

Another aspect of the importance of aroma is to be found in a thought-provoking book, *Delicious: The Evolution of Flavor and How It Made Us Human*, by Rob Dunn, a professor of applied biology, and Monica Sanchez, a medical anthropologist. They address the question of how flavour, which encompasses taste and aroma, has played a central role in the evolution of the human diet down through the ages. In so doing, they put aside the idea that our food choices were determined solely by availability and the need for certain nutrients. They speculate that the search for deliciousness and eating for pleasure rather than mere survival have more importance than is generally acknowledged.

Aroma family	How the aromas come about
Green	Produced when an oxidising enzyme (lipoxygenase) breaks down unsaturated fatty acids in cell membranes when the tissue is damaged. Examples: cucumbers and melons.
Fruity	Produced when enzymes in the fruit combine acid and alcohol molecules to produce an ester. Examples: apples, pears, and wine.
Terpene	Produced by enzymatically created substances that are characterised as having aromas that are flowery, citrusy, minty, herbaceous, and piney. Terpenes are very volatile and are the first aromas to disappear when heated. Examples: limonene in citrus fruits and menthol in mint.
Phenolic	Produced by enzymatically created substances, typically from an amino acid (phenylalanine). The substances are known from the smell of vanilla (vanillin), cloves (eugenol), thyme (thymol), and cinnamon (cinnamaldehyde). Phenols can bind to each other to produce polyphenols (tannins), which are bitter.
Sulfur	Produced enzymatically from tissue damage (for example, isothiocyanate). Examples: mustard, horseradish, cabbage, and onions.

Source: Based on McGee, H. *On Food and Cooking: The Science and Lore of the Kitchen*. Scribner, New York, 2004.

Red and green *shiso* (*Perilla frutescens*), an aromatic herb.

Type of aroma	Aromatics	Examples
Plants		
Freshly cut green leaves, grass	Alcohols, aldehydes with six carbon atoms	Most green vegetables, tomatoes, apples, other fruits
Cucumber	Alcohols, aldehydes with nine carbon atoms	Cucumbers, melons
Green vegetable	Pyrazines	Bell peppers, fresh peas
Earthy	Pyrazines, geosmin	Potatoes, beets
Cabbage-like	Sulfur compounds	Cabbage family
Oniony, mustardy	Sulfur compounds, for example, isothiocyanates	Onion family
Floral	Alcohols, terpenes, esters	Edible flowers
Edible mushrooms		
Mossy, earthy	Alcohols, aldehydes with eight carbon atoms	Button mushrooms, cremini
Musky	Many widely different substances	Truffles
Shiitake-*like*	Sulfur compounds (lenthionine)	*Shiitake*
Seaweeds		
Sea and fish smells	Iodine and sulfur compounds, for example, dimethyl sulfide	Brown algae, for example, sugar kelp, tangle, bladderwrack
Bromine-like, iodine-like	Halogenated compounds	Red algae
Marine	Halogenated compounds	Green algae
Floral	Unknown	Dulse

Source: McGee, H. *On Food and Cooking: The Science and Lore of the Kitchen.* Scribner, New York, 2004.

The Dorito effect—flavour based on additives

In his book *The Dorito Effect*, a Canadian journalist, Mark Schatzker, describes how the flavour of what we eat has been changing profoundly in the wake of the industrialisation of agriculture. The emphasis has shifted from the taste of the food to growing plants and raising animals as quickly and efficiently as possible and turning out uniform products that look attractive, are cheaper, and can be transported easily. The naturally occurring aromatics that should direct us toward healthy basic ingredients with the appropriate nutrient value have been sacrificed to meet these goals. Whole, fresh foods have become increasingly bland.

In order to compensate, an entire industry has grown up to develop and market an arsenal of aroma substances, both natural and synthetic, that are added to what we eat, even to what in an earlier time we would not even have thought of as 'food.' Doritos are a prime example. These chips were originally an ingenious and laudable way to avoid food waste by cutting up leftover corn tortillas, frying them, and adding some simple seasonings.

Since their humble beginnings in 1964, Doritos have become enormously popular, but have also been completely transformed, with a total of over 100 different varieties having been marketed around the world. This constant introduction of new flavours was, of course, made possible by using a vast number of additives, a technique that was lampooned in 1996 by a satirical magazine, *The Onion*, in an article headlined, "Doritos Celebrates One Millionth Ingredient."

According to Schatzker, this illustrates how our brain can be fooled into craving the wrong food. This is one of the reasons that a significant proportion of the population in many parts of the world is tending toward obesity and suffering from related lifestyle diseases that can be attributed to an unhealthy diet.

A similar development is already at work in the case of fake meat and meat substitutes. These are produced from plant protein by using a range of food additives to make these ultra-processed food products palatable, with questionable health effects.

Tomatoes are not like they used to be

For that matter, neither are most commercially grown strawberries, which seem to be twice as big as those in home gardens. This is a refrain that one hears over and over—they have little taste and they give off no enticing aromas. These impressions are not driven only by nostalgic feelings about how food was better in the olden days. It is certain that many fresh ingredients now taste very different from how their precursors did at the beginning of the twentieth century.

Why this is so is almost always explained by reference to the way in which our food production has become industrialised, turning out uniform produce, cheaply, and in great abundance. Plants are grown as rapidly as possible, often in a monoculture setting, with the aid of fertilisers and pesticides.

Apart from the production end, there have also been advances in selective breeding and hybridising techniques to develop new plant species and cultivars with traits such as good disease resistance and uniform ripening time, to obtain the greatest yield and the biggest profit in the shortest time.

Quality has come to be associated with appearance, availability, and nutritional value in products that are unvarying, often regardless of the season of the year. One might even question the extent to which the nutritional value is even factored into an assessment of quality.

But one thing seems to stand out: somewhere along the way, taste has gone missing in action. Part of the explanation for flavour's disappearing act is quite simple—the taste and aroma compounds are

now less concentrated. This is attributable to their being grown more quickly and to a larger size than in the past, leaving less opportunity for these substances to develop fully or in greater quantities. In addition, their cells are bloated with a higher water content.

Another, possibly more important, part of the explanation is that the strong focus on volume has allowed those species of plants that give a greater, faster rate of return to become dominant, at the expense of the less quantifiable criteria related to flavour. The best-known example of this approach is the ordinary supermarket tomato, which is grown in great abundance but is much less delicious than its heirloom cousin. Using genome analysis and flavour-mapping, scientists at the University of Florida have recently been able to determine that about one-third of the thirty most important taste and aroma compounds are present in significantly smaller quantities in commercial tomatoes than in the traditional varieties. Another team of international researchers has collected genetic information from 725 wild and cultivated tomatoes, with the aim of breeding back some of the desirable qualities that have been lost over the past half century. Their analysis also notably revealed a rare form of a flavour-imparting gene that is absent in most domesticated tomatoes. It depends on carotenoids, both to bring out the characteristic red colour of the ripe fruit and to make it taste delicious. As more plant breeders work to re-insert the missing genes to gain back their 'true' taste, one might dare to hope that the days of the pale, insipid mass-market tomatoes may slowly come to an end.

Tomatoes are not like they used to be

Irritation

The taste impression of some plants can be associated with irritation, or even outright pain. This sensation is due to the effect of certain substances in the plants on sensitive nerve endings in the mucous membranes of the oral cavity either by damaging the tissue or by binding to particular receptors on the cells. In all cases, it is a signal that tells us that food might harm us and that we should proceed with caution.

Vegetables such as horseradish, mustard, ginger, and onions or spices such as black pepper and chilli peppers all have irritants that affect the same nerve endings as those that tell us about the temperature of a food. They can trigger sensory impressions that are described as warm or burning, which is why we often refer to the 'taste' of chilli and pepper as being hot.

There is a variety of substances in plants that can provoke this impression. Black pepper and chillis contain piperine and capsaicin, respectively. In the case of horseradish, mustard, and *wasabi* the irritants are produced only when the cells of the plants are damaged, for example, by chopping or crushing, which causes certain enzymes to break down sulfur-containing compounds. As these enzymes are destroyed at higher temperatures, the burning and biting 'taste' disappears.

> **Cabbages—strong tastes and 'stinky' smells**
>
> The cabbage family includes many of the vegetables that appear most frequently on our plates. What they have in common is that they tend to taste bitter, they can have an irritating mouthfeel, and they generally give off a somewhat unpleasant odour. These are all due to their content of sulfur-containing compounds. But there is a big difference in the relative content of these substances, which is as follows: Brussels sprouts 35, kale 26, broccoli 17, green cabbage 15, red cabbage 10, radishes 7, Chinese cabbage 3, and cauliflower 2. If in doubt about eating members of the cabbage family, choose cauliflower, which is by far the mildest of the lot.

Purple cone cabbage.

Understanding the taste experience

Chilli peppers—from the mild to the smoking hot

Chilli peppers, which we can now find in an impressive array of shapes, tastes, colours, and degrees of spiciness, all belong to the genus *Capsicum*. They can all be traced to a single common ancestor in South America dating back almost seventeen million years. These wild plants evolved into three distinct groups centred on the coast of Brazil and in Andean areas and eventually made their way to other parts of the continent and to Central America. Their dispersal is commonly attributed to a symbiotic relationship with birds. Unlike mammals, birds do not have receptors in their mouths that respond to the burning taste of the peppers and the seeds pass through their digestive systems unharmed. Archaeological research has found starch microfossils dating back six thousand years at seven sites ranging from the Bahamas to Peru that are evidence of early domestication of chilli peppers.

Eventually humans took over the job that had been done by the birds, starting with Christopher Columbus more than 500 years ago. He was sailing west in search of a new spice trade route, specifically to find a way to import highly valuable peppercorns and other spices directly from India, when he instead bumped into an island in the Bahamas archipelago. While he did not find pepper, he was offered a spicy fruit by the natives. On account of its fiery flavour, he called it *pimiento*, after the Spanish name for pepper, *pimenta*, and took samples back to Europe. Although it did not catch on immediately in Europe, sea-faring Portuguese traders brought chilli pepper plants, probably originating in Brazil, to their settlements in West Africa, India, and many areas of East Asia, where they became enormously popular. Because they can be grown in abundance in a wide range of climates and are very affordable, chilli peppers are now estimated to be eaten by a quarter of the world's population every day all over the world. They can be considered the first truly globalised plants.

When we think about foods that are 'hot' we often associate them with chilli peppers, sometimes even describing them as being of 'industrial strength.' We know that this is due to their content of an irritant, capsaicin, but what determines how much of it there is in a given variety? Because chilli peppers feature so prominently in the food cultures of warm climates, there might be a tendency to conclude that those grown in sunny, hot places are more potent than those cultivated in more temperate zones. As it turns out, this is not the case—it can be shown scientifically that ripeness and the genetic inheritance of the plant are the most important factors. And it is also useful to pay attention to where most of the capsaicin is to be found in a chilli pepper. Contrary to popular belief, the seeds are not the culprits. Most of the capsaicin is contained in the whitish pith, called the placenta, in the middle of the fruit and its associated capsaicin glands to which the seeds are attached.

The first attempt to quantify the pungency of various types of chilli peppers was undertaken by Wilbur Scoville, an American pharmacist, in 1912. He used what is known an organoleptic test, which depends on sensory perception. It works by dissolving an exact weight of dried chilli pepper in alcohol to extract the capsaicin compounds (capsaicinoids), which are then diluted in sugar water. Samples of this solution, with decreasing concentrations of the capsaicinoids, are evaluated by a panel of five trained testers until their presence can no longer be detected. The heat level is expressed in Scoville Heat Units (SHU), which are rated in multiples of 100. An SHU is defined as follows: If a chilli pepper extract has a value of 1,000 units it has to be diluted 1,000 times before it can no longer be detected. Unfortunately, the Scoville method is rather imprecise as it depends on subjective impressions and the number of mouth heat receptors, which vary from one individual to another, and the sensory

Chilli peppers can be mild and sweet, but most are quite spicy, while others are almost too hot to handle.

fatigue that sets in when a tester's palate is exposed to a few samples in rapid succession. This way of analysing the pungency of a chilli pepper has now largely been supplanted by a chemical laboratory method, called high-performance liquid chromatography. Using this precise tool, the strength of pure capsaicin is defined as 16,100,000 SHU.

At present there are twenty-six known wild species of chilli peppers, but most of us are familiar only with the more popular cultivars of the five domesticated species that are grown commercially. But sub-species continue to proliferate as plant breeders create new ones in order to optimise desirable traits in the plants or, in some cases, to try to come up with ones that have an even hotter and record-setting SHU rating.

How hot are chilli peppers? The Carolina Reaper, a hybrid plant that also has sweet and fruity undertones, has been tested at 1,600,000 to 2,200,000 SHU and was until recently considered the world's hottest. It was dethroned in 2023 by Pepper X, which has a rating of 2,700,000 SHU. Various types of habañeros have a pungency rating of between 100,000 and 750,000 SHU, while the Thai varieties measure between 50,000 and 100,000 SHU. Much milder ones are serranos (10,000 to 25,000 SHU), jalapeños (2,500 to 10,000 SHU), and poblanos (1,000 to 2,500 SHU). Cherry peppers, Italian pepperoncini, and banana peppers are all very mild, with ratings of from 0 to 500 SHU. Due to a recessive gene, ordinary bell peppers do not produce capsaicin and are rated 0 on the Scoville scale.

Perhaps one should leave the last word on why chilli peppers are so central in many cuisines to Madhur Jaffrey, the renowned Indian food writer. She says that once eaten, a chilli pepper "provides a high, there is no going back. It turns into a craving. The chilli is not so much a seed of change, as a conqueror, or better still, a master seducer."

Astringency

Many unripe fruits cause the mucous membranes to pucker up in an unpleasant way, a taste experience described as astringency. This is a signal from the plant that the fruits are not yet ready to be eaten. Well-known examples are unripe persimmons and bananas that are still a bit green, as well as the skin from grapes and some seeds, for example, caraway and fennel. Vegetables that have many red pigments, such as the anthocyanins in radicchio and red-leaf lettuces, may also evoke this sensation because of their polyphenol content.

In order to counter the effect of astringency, one can try to suppress the mechanism that causes it, namely the chemical reactions between polyphenols in the food and the surface of the tongue and saliva. The polyphenols bind with proteins in the saliva, so that they clump together and make it more difficult for the food to slide over and past the mucous membranes in the mouth. Acids and salt amplify astringency. Sugar, dairy fats, proteins such as those in gelatine, and polysaccharides that bind water (pectin and other hydrogels) have the opposite effect because they bind with the polyphenols before they have a chance to grab on to the proteins in the saliva.

Texture

A very important sensory property of plant-based food is its mouthfeel, more commonly called texture. It is often what determines whether or not we like a particular vegetable. The classic example is the difference between vegetables that have been cooked to death so that they are mushy compared with how we react to them if they are lightly steamed or even raw, leaving them crisp and juicy.

The texture of vegetables is dependent on the structure of the plant at the cellular level, how the cells are bound together by different types of carbohydrates, and even the nature of the cell walls. It also depends to a great extent on the turgor pressure inside the cells.

The carbohydrate make-up of a particular plant has the greatest influence on how we experience its mouthfeel. Some carbohydrates, such as cellulose, are essential building components that help to keep the plant upright and stiff. Others, for example, pectin and hemicellulose, are used to bind the cells together and strengthen the cellulose content of the cell walls. The relationship between the way different carbohydrates bind to each other and their varying ability to bind water results in a whole range of textures—from firm, hard, crisp, soft, juicy, and crunchy to tough, dry, and mealy.

Some parts of the plants, especially green leaves, incorporate a significant amount of air, up to 70 per cent, which helps them to spread out to absorb sunlight. This is why they collapse when they are cooked. Exceptions to this are the vegetables, such as leeks and onions, that are made up of swollen scale-like leaves that are packed tightly into a bulb to store up the nutrients that tide them over the non-growing season. These keep their shape after they have been heated.

The starch content of root vegetables varies widely, with potatoes having a great deal of it, while others, such as carrots, turnips, and beets, have very little. As starch absorbs water, cooked potatoes feel very soft, but also dry and mealy, whereas a carrot can remain juicy and firm.

In the kitchen, the application of heat is the most common way of altering the texture of plants. Marinating with salt, acid, and sugar affects the structure of the various carbohydrates such as pectin and hemicellulose and is also used to change texture. We will return to that later.

There are two simple possibilities for turning cooked, mealy starchy vegetables into something more appetising. One is to add fats, for example, mashing potatoes with butter. The other is to introduce a texture contrast by combining a serving of mealy ingredients with some very crisp ones, such as *tsukemono* (see *The Flavour Accelerators*). Alternatively, one can add crunch using a variety of toppings, including lightly toasted nuts, seeds, croutons, and fried bacon bits.

Unripe fruits and berries can also be utilised as a sort of vegetable as long as they are not too sour, bitter, or astringent. Examples are very firm plums, as well as pears, apples, mangos, and even strawberries that are still green. Some types of unripe fruit, especially plums, can be pickled using just salt and end up with a crisp and sometimes crunchy texture.

Unripe plums pickled in salt are crisp and crunchy.

Fresh vegetables and fruits can be quite crisp and juicy. As they ripen, the bonds between the stiff cell walls weaken, for example, when pectin is broken down by enzymes in overripe apples. Their texture becomes mealy and dry and can feel a little grainy. This is not necessarily due to their drying out as they can contain just as much juice as a crisp apple.

In an apple that is not fully ripe, the cells are tightly bound to each other. When one bites into it, the bonds are so strong that they hold, and it is instead the cells themselves that break apart. This is why the apple feels crisp and crunches between the teeth. Because the juice comes pouring out of the cells, we describe the apple as being juicy. But when we bite into an overripe apple, the cells are more loosely bound and will slide past each other without bursting. Its mouthfeel will then probably be characterised as soft, dry, and possibly mealy.

Sauce tomate—the fifth mother sauce

Shortly before his death, the celebrated French chef Marie-Antoine Carême (1784-1833) proposed a formal classification of the basic or 'mother' sauces, dividing them into four categories: *espagnole*, *velouté*, *allemande*, and *béchamel*. In 1903, an equally famous French chef, August Escoffier (1846-1935), revised Carême's work, demoting *sauce allemande* to the status of a 'daughter' sauce of *velouté* and adding *sauce hollandaise* and *sauce tomate*. Unlike the others, *sauce tomate* was based on a vegetable, tomatoes, rather than stock, milk, or eggs and butter. Although this was considered an innovation in French cuisine, sauces made with tomatoes were already popular in Central America at the time of the Spanish conquest early in the sixteenth century. In the local Nahuatl language, the smaller green and sour tomatoes were called *tomatl*, while the larger red ones were known as *xitomatl*.

The tomato pulp serves both to add umami taste and to thicken the sauce. Escoffier's original recipe calls for crushed raw tomatoes, but most modern versions use cooked, peeled, and chopped tomatoes from a can. Within the taxonomy of classic sauces, *sauce tomate* also has a number of 'daughter' sauces, for example, ketchup, sauce Bolognese, barbeque sauce, and marinara. In principle when it is made with tomatoes, even *tikka masala* sauce could also be included in their ranks.

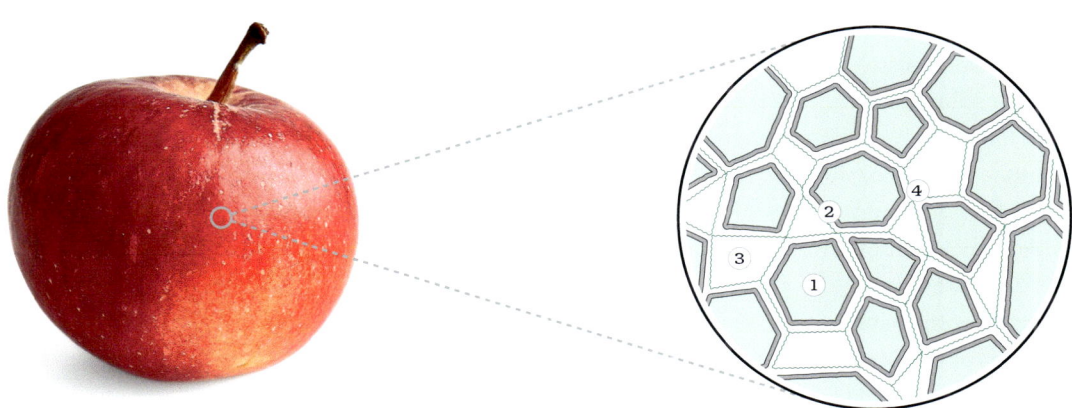

The structure of cells in an apple: Juice (1), cell wall (2), air pocket (3), and pectin bonds (4).

The taste of plants

The taste of mushrooms

Not all mushrooms are created equal. Some are inedible, in fact poisonous, and others are unpalatable. Fortunately, the majority are edible, although only about twenty species have become culinary staples. The more popular ones have a characteristic rich earthy and meaty taste. This is due to their having not only a fairly substantial amount of free glutamate, which contributes basal umami, but also an abundance of free nucleotides, especially guanylate, that in combination with the glutamate creates synergistic umami. As a result, edible mushrooms make ideal partners for vegetables, which are generally lacking in umami.

It is important to choose those mushrooms that are the best sources of umami. The rule of thumb is that in all cases dried ones have more guanylate than when they are fresh because free nucleotides develop during the drying process. Also, the darkest coloured mushrooms are better sources of umami. Truffles are an exception, though, as the white ones have more umami than the black ones.

Recently it has been discovered that some mushrooms, such as porcini mushrooms (*Boletus edulis*), are a good source of the substance glutathione, a tripeptide, known to elicit the sensation of *koku*.

The most characteristic smell of mushrooms is due to 3-octanol, a liquid substance that is formed when their tissue is damaged, allowing certain enzymes to break down their polyunsaturated fats. While this liquid is tasteless, it is very aromatic, having a mossy and earthy smell, with notes of nuts and citrus fruits. In large quantities it has a somewhat metallic odour.

Although mushrooms are largely made up of water—between 80 and 90 per cent—they retain much of it because their cells are protected by an outer layer of hard, insoluble chitin. This polysaccharide strengthens the cells and ensures that the mushrooms stay firm rather than becoming soft and mushy when they undergo a lengthy preparation. Heating mushrooms tends to make them firmer because they shrink when they lose water, and the air pockets collapse.

Gilled mushrooms have an especially large content of the types of enzymes that result in browning reactions. This can be seen clearly in the case of white button mushrooms, which turn brown when they are damaged or bruised. This also results in the formation of water-soluble dark substances that can discolour other ingredients in a dish to which the mushrooms are added. Browning can be avoided by heating the mushrooms or exposing them to acidic or basic substance to weaken the enzymes.

The giant puffball (*Calvatia gigantea*) is easy to identify from its round fruiting body and lack of a stalk. It has a mild taste and aroma and is edible when young provided its interior is still white and firm. This makes it an ideal candidate for pan-frying. Unfortunately, puffballs are not farmed commercially, so one would have to depend on foraging to obtain them.

Shiitake take the prize

Shiitake (*Lentinus edodes*) are the classical example of mushrooms that have much more guanylate after they have been dried. These mushrooms are very closely connected to the history of umami, because it was Japanese Zen monks who discovered that they could make a *dashi* that satisfied their vegan dietary requirements by using dried *shiitake* instead of *katsuobushi*, a special type of dried fish. (See The Flavour Accelerators.) Fresh *shiitake* have a woody and slightly sour taste but develop a whole array of taste and aroma substances when they are dehydrated. Among them is a special sulfur-containing compound, lenthionine, which is formed in the gills, formally known as lamellae, by the action of the mushrooms' own enzymes. This same process occurs when the dried *shiitake* are soaked in lukewarm water. In order not to weaken the action of the enzymes, one should avoid soaking dried *shiitake* in water that is too warm or cooking or grilling them. *Donko* (meaning 'winter mushroom') is a high-grade variety of *shiitake* that matures over a long period of time, allowing it to develop an abundance of guanylate and umami. Even though *shiitake* are very rich in guanylate, other mushrooms, for example, the ordinary button mushroom (*Agaricus bisporus*) and chestnut mushroom (*Agrocybe aegerita*), can elicit more umami taste by a combination of free glutamate and free nucleotides.

Truffles—in a category by themselves

Truffles are no mere mushrooms. Like tomatoes, they are in a category by themselves, particularly those from the two families that are native to Europe, black truffles (*Tuber melanosporum*) and white truffles (*Tuber magnatum*). Their characteristic tastes and aromas are so pungent that they are considered to be an experience that no gourmet should miss. Along with caviar, saffron, and *matsutake* mushrooms, they are among the most expensive delicacies in the world. In order to understand how this came about, we need to know a little about how they grow and their chequered history.

How they grow
Truffles grow best in chalky soil under oak, beech, and hazelnut trees that are at least six years old, with which they establish a symbiotic relationship called a mycorrhiza. Like other mushrooms, their base is an underground mycelium network made up of tiny hair-like threads that wrap themselves around the roots of the trees and through which they take in sugar and other nutrients. In return, the mycelium network absorbs water and minerals for the use of the trees, greatly expanding the reach of their root system. But unlike other mushrooms, when the mycelium sets fruiting bodies, they remain as far as half a meter below the surface. It takes about six months for the truffles to reach maturity at which time they look more or less like ugly, irregularly shaped spheres, ranging in size from that of a walnut to that of a large orange. In order to reproduce, they depend on either animals or humans to dig them up and disperse their spores.

Truffles have been eaten since ancient times
Foraging for truffles growing naturally in forests goes back at least four thousand years. Because of their seemingly mysterious growth habit, so different from that of other edible plants, they were the subject of myth and much speculation in the ancient world. Early Greek philosophers and writers were puzzled about their origins and suggested they could be attributed to divine intervention or grew where thunderbolts hit the ground. Truffles were praised as being seedlings of the gods, while others rather suspiciously wondered whether they were warts on the skin of the earth or infants belonging to sorcerers and witches. They were thought to have both medicinal and aphrodisiac properties. The Romans more prosaically regarded them simply as *Tuber terrae*, fruits of the earth, but considered them a decadent indulgence, with a price to match. In *De re coquinaria* (On the Subject of Cooking), the cookbook attributed to the first-century Roman gourmet Apicius, one finds no less than five recipes for truffles in the section devoted to sumptuous dishes.

These delicious mushrooms fell out of favour in Europe during the Middle Ages partly because religious authorities could not explain them according to the system of natural laws that dictated what people should eat. Foods that grew wild and in the ground were already deemed to be less desirable. Worse yet, because the fruiting bodies of truffles grow completely underground, they were suspected of having satanic origins, to the extent that sermons were preached against them. Nevertheless, after they were served at the papal court in Avignon in the fourteenth century, they soon gained the approval of church authorities as permissible pleasures. It did not take long for them to start to grace the tables of the noble and the wealthy, including King Francis I of France and Catherine de Medici. From that time onward, they have been one of the most treasured ingredients in *haute cuisine*, elevated to the status of "the black diamonds of the kitchen" by Brillat-Savarin.

As might be expected, a product that was expensive and held in such high esteem started to attract the attention of would-be growers. The first known attempt at *trufficulture*, as it is known in France, goes back to about 1790. A miller by the name of Paul Mauléon living in the Val de Loire region, who was a keen observer of nature, made the connection between oak trees and truffles. He planted acorns gathered from trees where truffles had been found and planted them in similar conditions in a secret location. Several years later truffles were found

Ravioli with black truffles.

around these new oak trees. In a similar experiment in the mid 1800s, Auguste Rousseau planted a relatively large stand of oaks that yielded a large harvest of truffles, an event that was widely publicised.

Trufficulture soon spread to other parts of France and Italy, partly given a hand by the epidemic of phylloxera that destroyed many vineyards in southern France, followed by a disease that killed silkworms, rendering the mulberry trees on which they fed useless. Because the soil in these fields could also support the farming of truffles, it became an important source of income in these areas. By the end of the century so many truffles were being dug up that they were sold at farmers markets and were eaten by all social classes. Once again, however, a series of external events resulted in some drastic changes that led to a steep drop-off in truffle production. Agricultural workers left to fight wars and industrialisation drew people from rural areas to the cities, taking with them valuable knowledge about truffle cultivation. Plantations became overgrown because there were fewer sheep grazing there and trees were either not pruned or used for firewood. Farming truffles was notoriously tricky and those who stayed the course had to deal with the vagaries of weather, precipitation, and other factors that introduced uncertainty into the operation. Inevitably, the price of truffles increased and they regained their status as a luxury food.

Perhaps there is hope that the price of truffles will come down at some point. In the last half century researchers have been developing more reliable ways to establish mycorrhizas. Money is being invested to establish carefully cultivated plantations using new techniques that will mitigate some problems such as contamination from other species. Experts say that there is no difference in taste and aroma between cultivated and wild specimens. About 80 per cent of the truffle production in France is now farmed, compensating somewhat for the decline in wild ones. Because the prices are still exorbitantly high, the truffle groves are often well hidden to prevent theft by freelance 'foragers.' There are generally also very strict rules regarding where truffles can be hunted on public lands and who is permitted to do so.

Hunting for truffles

It may seem a little incongruous to speak of 'hunting' for truffles rather than 'harvesting' them. But given how it is done it is an especially apt choice of word and it is a labour-intensive activity. As they are not on the surface, the truffles have first to be located. One way to do so is to look for bare patches or dry grass near the trees where the fungi have absorbed the nutrients in the soil for themselves, to the detriment of other vegetation. Another way depends on detecting the aromatic substances with a very strong musky smell that is given off when the truffle is ripe. At one time the farmers and foragers used sows, with their highly developed sense of smell, to search for the fungi, but it was very tricky because the pigs liked to devour them immediately. Also, when they rooted around in the ground to dig them out, they destroyed the mycelium network, leading to a decrease in fruiting bodies. As a consequence, the use of truffle pigs has declined dramatically and is outright banned in some places. The work has now been passed on to specially trained dogs. While they are still puppies they are paired with older, experienced truffle hounds and learn how to smell out the truffles. They are also taught that they are not allowed to eat them. Luckily it is much easier to distract a dog from snacking on the fungus by giving it a treat than it is to wrestle it away from a 150-kilogram pig.

The hunters and their dogs may cover the same ground day after day because the tell-tale odour that is the sign of ripening literally happens overnight. Sometimes there is not so much as a whiff of truffle aroma one day, but the next it might be ever so pungent. When one is found, it can be up to a half metre down. It is dug up very gently with a special tool that resembles a cross between a knife and a trowel. The truffle is then rinsed, the dirt is scrubbed off, and it is allowed to dry. A good one should feel firm but be slightly elastic and give a little when pressed. It will keep for about a week.

> *"Presently, we were aware of an odour gradually coming towards us, something musky, fiery, savoury, mysterious—a hot drowsy smell, that lulls the senses, and yet inflames them—the truffles were coming."*
>
> William Makepeace Thackeray, *Memorials of Gourmandizing*, 1841

Tastes and aromas—positively addictive

It is very easy to justify a love of truffles on a nutritional basis because they have few calories and an abundance of important vitamins and minerals. But they are an indulgence and should be handled carefully for maximum enjoyment. They really come into their own when paired with simple ingredients that do not compete with the umami and the sensations of *koku* for which they are famous.

When the black truffles are ripe and ready to disperse their spores, they emit a very strong, musky aroma, made up of a wide range of scent molecules. One of these is anandamide, a chemical similar to the psychoactive compound in marijuana, known as the 'bliss molecule,' for its ability to trigger feelings of happiness. Is it any wonder that people think of eating truffles as an unforgettable experience? Their taste is earthy and a little sulfurous, thanks to the presence of dimethyl sulfide and has notes of sweet fruits. They lose their aroma when they are heated and consequently are added to a warm dish only at the very end.

Ripe white truffles also have a complex aroma profile described as having notes of honey, hay, garlic, spices, wet earth, and ammonia. They have a stronger, sharper, and more bitter taste than their black cousins that is, nevertheless, rather subtle. As a result, they are more sought after and much more expensive. They lose their aroma very quickly and in order to preserve it they are also often left raw and shaved or sliced over a fully cooked dish just before it is to be eaten.

Reflecting on the almost mystical appeal of truffles brings to mind the beloved nineteenth-century Italian composer Gioachino Rossini, who is best known for his thirty-nine operas. But he was also a renowned gourmet and was exceptionally fond of truffles, which he described as "the Mozart of mushrooms." He once said, "I have wept three times in my life. Once when my first opera failed. Once again, the first time I heard Paganini play the violin. And once when a truffled turkey fell overboard at a boating picnic."

Seaweeds—the taste and smell of the sea

Like plants, there are so many different types of edible seaweeds (macroalga) that it is difficult to characterise unambiguously the overall taste experience associated with these algae. But in general, it engages all five senses. Their textures are especially prized in Asian cuisines and some species of seaweeds are valued for contributing aesthetic appeal due to their colours, shapes, and patterns.

Konbu (Saccharina japonica) is a brown alga with an intense umami taste.

Aroma

The aroma of seaweeds is often undervalued because foul smelling breakdown products containing sulfur can be present in large quantities. But in small amounts, substances such as dimethyl sulfide emit odours that remind us of a pleasant sea breeze, which often also has distinct notes of iodine and bromine.

One of the most popular seaweeds is the Japanese *konbu* (*Saccharina japonica*), which is rich in umami and is a primary ingredient in the soup stock *dashi*. After the seaweeds have been harvested and dried, they are aged in temperature- and humidity-controlled environments for at least two years. Those of the highest quality may even be aged for ten years or more. During this time, the intense smell of the sea and unpalatable substances are broken down, with the result that the natural umami becomes more pronounced.

In other parts of the world seaweed harvesters dry some types of brown algae, such as the giant kelps (*Macrocystis pyrifera*) and bull whip kelp (*Nereocystis luetkeana*), in the sun. The ultraviolet rays of the sun break down bitter-tasting polyphenols.

Taste

Umami is undoubtedly the most prominent taste associated with some species of seaweeds thanks to their free glutamate content, which varies quite widely among the different types. Some species of Japanese *konbu* have such an abundance of it that it places them in a special category, which is directly reflected in the price that they fetch. But a red alga, dulse (*Palmaria palmata*), has as much glutamate as some of the less-prized species of *konbu*. Other types of seaweeds that have a reasonable glutamate content are *wakame* (*Undaria pinnatifida*) and laver (*nori*, *Porphyra/Pyropia* spp.). On the other hand, most species of brown algae, for example, sugar kelp, tangle, and bladderwrack, have very little glutamate.

Recent research has shown that there is not a simple relationship between how a person experiences the taste of umami in seaweeds and the glutamate content. It appears that there is a multisensory interaction with certain aromatic substances in the seaweeds that are perceived as umami.

Laver (*Palmaria palmata*) is a species of red alga with an abundance of umami.

Seaweeds prepared three ways: marinated in vinegar, simmered in soy sauce (*konbu tsukedani*), and simmered in soy sauce and sprinkled with soy powder.

Texture and mouthfeel

Even though umami taste and the special smell of the sea are what we most often associate with seaweeds, they are actually more sought after for their contribution to the mouthfeel of a dish in those food cultures where they are highly valued. The many different species that find their way into the kitchen exhibit a broad range of textures, which depend on whether the seaweeds are young or mature specimens, which parts of the alga are being used, how long they have been aged after harvesting, and not least how they are being prepared.

Apart from very young shoots, most species of seaweeds are rather tough. Even the very thin and delicate fronds of sea lettuce (*Ulva lactuca*) and laver (*Porphyra/Pyropia* spp.) can be quite chewy. Drying, toasting, and cooking tenderise them somewhat. The larger seaweeds, for example, *wakame* (*Undaria pinnatifida*) and winged kelp (*Alaria esculenta*), have a more differentiated tissue structure with midribs, which makes them tougher. Other species have stems and air bladders that can be very difficult to chew and, in many cases, are inedible. The parts of the fronds that house the sporophylls on species such as *wakame* and winged kelp have a significant fat content and, as a result, are often more delicate and tastier than the rest of the seaweed.

The enzymes in dried seaweeds can remain active while they are being aged and thereby contribute to the tenderising process, which can also be speeded up by rehydrating the dried seaweed quickly. This is done, for example, in the case of the red alga laver (*Palmaria palmata*) so that it ends up with an almost soft liquorice-like texture.

Seaweed novelties

Dried, toasted, and deep-fried seaweeds, whether green, red, or brown species, are much appreciated as crisp snacks. Like fruit, they can also be candied and eaten as is or in a dessert. Toasted and granulated or powdered seaweeds are often used as taste additives or as salt substitutes in a whole range of foods, such as bread, cheese, and sausages.

Crispy 'seaweed bacon' made with the red alga dulse (*Palmaria palmata*) was recently launched in the United States with great fanfare as a special new product. But a simpler version has actually been around for a long time in Scotland and Ireland where it earlier was served as a snack in the pubs. It is quite a simple idea—dried, and sometimes also smoked, dulse fronds are deep-fried. The resulting texture is similar to that of regular bacon made with pork and when smoked correctly this 'seaweed bacon' can easily be used as a substitute in dishes such as omelettes and mashed potatoes. As an added bonus, it is a very low-fat product.

Some brown algae, notably sugar kelp (*Saccharina latissima*), exude large quantities of certain polysaccharides, such as alginate, as well as the sweet substance mannitol. Even though alginates are quite useful in the kitchen as gelation agents, this can be a drawback if these species are incorporated into a salad or a soup because they can result in a slimy mouthfeel or excess viscosity. Some Japanese seaweed products, for example, *konbu tsukedani*, which is *konbu* simmered in soy sauce, has a strong umami taste and a soft, but firm, texture, almost like a soft liquorice (see *The Flavour Accelerators*).

Many commercial seaweed products are prepared to emphasise crispness and crunchiness, in that way also engaging our sense of hearing as we bite into them. The best-known example is probably *nori*, which is a toasted sheet of seaweed paper made from the red alga laver (*Porphyra*/*Pyropia* spp.). This is used to make sushi rolls or crushed into flakes in the form of *furikake* that can be sprinkled on salads and cooked rice. As dried seaweeds readily absorb moisture from the air or from other food ingredients, products made with *nori* should be consumed quickly before they become soggy. Consequently, sushi rolls that are wrapped in *nori* (*hoso-maki*) should be eaten as soon as they are prepared in order to enjoy fully the contrast between their crisp exterior and the soft rice interior.

The fronds of winged kelp and *wakame* have a midrib, much like the leaves of many plants, for example, kale. It has a rather firm texture and can be deep-fried to make a delicious snack.

Bladderwrack in a dish with wild beach cabbage and blue mussels.

Seaweeds—the taste and smell of the sea

GREEN CUISINE COMES IN MANY COLOURS

the importance of eye appeal

Like aroma, colour can have an important influence on how we react to a plant-based dish even before we have had a chance to taste it. Vegetables, fruits, seaweeds, and fungi owe their visual appearance to chlorophyll and three other types of pigments that give rise to all the colours of the rainbow, as well as white, brown, and black. Knowing a little about the properties of these pigments and how they react to heat, acidity, and so on can be used to great advantage to enhance the visual appeal and preserve the nutritional value of green cuisine.

The colour of vegetables, fruits, fungi, and seaweeds can undergo changes in the course of their lifecycle as they grow and ripen, depending on exposure to sunlight, and with the changing seasons. Colour is also affected by how they are prepared for eating. For example, the balance between the different pigments can be disrupted when otherwise dominant pigments with colours such as brown or purple are broken down and green chlorophyll becomes prominent. This effect can be seen when purple beans are blanched and turn completely green.

One can learn much by paying attention to colour when one is evaluating whether any of these living ingredients are either sufficiently ripe or, in order to avoid wasting food, still fresh enough to eat. This applies especially to changes that occur naturally as the organisms age, for example, when they fade or turn brown. There can also be a question of different chemical reactions that lead to bruising, either because the ingredients were damaged in shipping and storage or when being prepared in the kitchen.

Four types of pigments

Chlorophyll, which is the source of the dominant green colour of so many plants, is probably the pigment with which we are most familiar. The other pigment types are the flavonoids, carotenoids, and betalains, which are responsible for red, orange, yellow, blue, and purple hues. These can protect the plants from harmful ultraviolet rays and in a nutritional context can be a source of important antioxidants.

Chlorophyll: green

Chlorophyll is a natural compound that is vital to

the process of photosynthesis in plants and algae by helping them absorb energy from sunlight to convert water, nutrients, and minerals from the soil, and carbon dioxide from the air into carbohydrates and to give off oxygen.

There are two types of chlorophyll: chlorophyll *a* has a strong green colour with a bluish sheen and is found primarily in plants exposed to a great deal of sunlight; chlorophyll *b* has a less intense olive colour. Both types are soluble in oil and are insoluble in water. Normally chlorophyll *a* predominates in the leaves of the plants but the balance can shift toward chlorophyll *b* in plants that grow in the shade or are aeging. As heat and acidity can have a strong effect on chlorophyll *a*, it is important to take this into account if one wants to preserve the green colour of certain plant-based foods.

The chlorophyll molecule can also become soluble and seep into the cooking water as a result of an enzymatic process or in the presence of acids and bases. In many cases the acids are released from the cells of the plant itself when they are injured during preparation. When this causes the acidity level of the water to become too high, the magnesium atom in the middle of the chlorophyll can be displaced by a hydrogen atom and the vegetables lose colour.

This is why the golden rule is to handle green ingredients very gently by heating them quickly in an abundance of water, thereby diluting the acids that have seeped out of the plants. If one needs to add some acidity such as lemon juice or vinegar to a vegetable dish or a salad, it is best to do so just before serving.

Flavonoids: red, blue, purple, and yellow
Apart from having powerful antioxidant properties and being an excellent source of vital phytonutrients found in almost all fruits and vegetables, the flavonoids are responsible for their vivid colours. There are a number of significant groups of flavonoids, each of which has subgroups originating in different foods, among them anthocyanins and anthoxanthins.

Anthocyanins are the source of the red, blue, and purple colours of foods such as blueberries, radishes, red cabbage, and purple cultivars of asparagus and beans. These pigments are water soluble and can seep out into the cooking water. They are probably also best known for their ability to change colour, depending on the acidity of their environment. When it is sour or acidic, they are blue and when it is basic, they take on red hues. One can take advantage of this reaction to obtain a particular colour, for example, by pickling. Anthocyanins are also oxidised by contact with metals, which can sometimes lead to unwanted colour changes.

Anthoxanthins are a related group of pigments that bring out yellow tones and are found, for example, in cauliflower, potatoes, and onions. They are also water soluble and the intensity of their colour fades when such vegetables are cooked in water.

Blanching vegetables

Blanching involves plunging the vegetables for a short time into boiling water. This renders the enzymes that could lead to browning quite harmless and causes the air pockets that are found in the spaces between the cells in the plant tissue to collapse. As a result, their green colour becomes more intense and pronounced. Immediately after blanching the vegetables are normally immersed in very cold water in order to stop the cooking process and preserve their crisp texture. Apart from that, one avoids damaging the chlorophyll by exposure to high heat for a long period of time. Some plants, for example green beans and some members of the cabbage family, take on a clearer, stronger green colour because blanching causes their red and purple pigments to seep out into the water. This effect is seen very clearly when purple kale and purple beans are blanched and, perhaps somewhat disappointingly, lose some of their eye appeal as they turn out to look like their more common green cousins.

Purple kohlrabi, whose colour is due to anthocyanins.

Carotenoids: red, yellow, and orange
The carotenoids are a family of fairly large pigment molecules in shades of red, yellow, and orange. These substances enhance the overall efficiency of the photosynthesis but also protect the photosynthetic system from harms caused by excessive exposure to light. The yellow ones are found, for example, in mangos and sweet potatoes, the orange ones in carrots and yams, and the red ones in watermelon and tomatoes. A number of health benefits are associated with incorporating fruits and vegetables rich in carotenoids into our diet as they have antioxidant properties. In particular, those that are brightly coloured are rich in beta-carotene, which is broken down in the body to produce vitamin A.

Like chlorophyll, carotenoids are soluble in fats, but not in water, and consequently, do not seep out into cooking water. They are more stable and are less affected by heat and acidity than chlorophyll and anthocyanins. This is why the carrots lose only a little of their orange colour when they are cooked or pickled in vinegar or juice from citrus fruits.

Betalains: red and yellow
Betalains are found in only a few plant families and are made up of two subgroups, the approximately fifty red betacyanins and twenty or so yellow betaxanthins. They are responsible for many possible colour combinations, for example, the red and yellow colours of beets and prickly pears. Betalains are of commercial importance as a source of natural food dyes and recent studies have indicated that they have potential as antioxidants. Their colours are affected by heat and are water soluble, which is why cooking water from beets turns red. About one in ten people are unable to break down the red pigments in the course of digestion and excrete them, making their urine and faeces look red after eating beets.

Carotenoids are responsible for the orange colour of the most common varieties of carrots.

Four types of pigments

Candy cane beets

Carrots and beets store up their carbohydrate energy depots in a special way using vascular tissues. Apart from providing mechanical support, these tissues serve as a transport corridor. It consists of two systems, the xylem that moves water and minerals upward to the leaves and the phloem that transports the sugars that are the products of photosynthesis back down to be stored in, and build up, the root. Xylem and phloem have differing abilities to dissolve various pigments and, as a result, the latter often have a more intense colour and one may be able to differentiate between the two when looking at the vegetables in cross-section. In carrots the xylem are bundled together in a central inner core that runs the length of the root. It is generally lighter in colour (p. 73) and has a somewhat tougher texture than the stored-up nutrients that completely surround it. The phloem deposits the carbohydrates in this outer part of the carrot and it, therefore, tastes sweeter. Beets, on the other hand, are built up of concentric alternating layers of these two types of tissues. Again, because of the differing abilities of the xylem and phloem to dissolve pigments it is usually possible to detect the ring-like pattern that is formed. In a special variety, called candy cane beets, also known as Chioggia beets, this effect is especially pronounced.

Cross section of a candy cane beet.

The colours of mushrooms

As mushrooms lack chlorophyll *a*, they do not photosynthesise and must derive their energy from the plants with which they have a symbiotic or a parasitic relationship. Nevertheless, they are able to obtain sufficient nutrients to biosynthesise an extraordinary range of pigments. Their colours and the ability to change them are really a tool that they use to discourage animals either from eating them before their spores are ripe or to do so when the spores are ready to be dispersed.

The fruiting body and the stalk of the cultivated ones with which we are most familiar commonly range from chalk white through shades of brown to almost black. Others, generally ones that grow in the wild and that may not be edible, also exhibit vivid blue, red, and dark green hues. Some change colour, for example, from white to brown, red, or blue, when they are cut into or otherwise damaged. Like plants and fruits, pale mushrooms can turn brown due to enzymatic browning.

Browning

There is a wide range of chemical reactions that can cause vegetables, fruits, and mushrooms to turn brown. In some cases, the results are undesirable, while in others they have positive outcomes.

Browning that is caused either by enzymatic or non-enzymatic action is generally to be avoided. It can have a negative aesthetic impact and can also result in bitter tastes. It usually occurs after the vegetable or fruit has been bruised or cut into pieces.

Conversely, browning can take place in a controlled fashion in the kitchen with the use of heat. Examples are caramelising using sugar or by high-temperature roasting or sautéing to generate Maillard reactions. Both of these methods lead to appetising taste and aromatic substances that we find very palatable.

Enzymatic browning
Enzymatic browning is a chemical reaction that takes place in the presence of oxygen when an enzyme called polyphenol oxidase (PPO), or another enzyme, converts the plant phenols to polyphenols, which are brown and can be bitter and astringent. These enzymes are found in the plant cells but have no access to the phenols unless the plant is damaged or cut open. Examples of this effect are when apples and celeriac turn brown sometime after they have been sliced into pieces.

One can prevent browning by either blanching to destroy the enzymes, excluding oxygen from the cut surfaces, which can be difficult to do in the kitchen, or weakening the enzymes by creating conditions that are either more acidic or more basic (pH < 5.0 or pH > 7.5). For example, apple slices can be sprinkled with lemon juice. As the enzymes that cause browning work more slowly at lower temperatures, one can also slow down the process by placing the vegetables and fruits in very cold water or putting them in the refrigerator. The advantage of using cold water is that this minimises exposure to the air. Sulfites, which bind to the phenols, also serve to reduce the amount of browning. This effect is used to advantage to preserve the natural colours of dried fruits.

Non-enzymatic browning
Fruits and fruit products, including juices and wine, that contain ascorbic acid (vitamin C) can turn brown, even without any help from enzymatic activity. The brown substances formed often have an unpleasant, bitter taste. This can occur in the absence of oxygen, but it happens more quickly when it is present. In order to counter this type of browning, one can add sulfites to prevent oxidation of the ascorbic acid, as is often done in the case of wine and beer.

Caramelisation
Caramelisation takes place when various sugars that are present in the food are broken down by heat and bind to each other to create tasty brown compounds in a sticky or solid mass that we usually refer to as caramel. This polymerisation process creates both large, long molecules (caramel substances) and some smaller, volatile molecules that give off the special aromas—fruity, nutty, toasty, creamy, and buttery—that we associate with the process.

For the majority of sugars, caramelisation takes place at temperatures of 160–180°C (320–355°F) and at slightly lower ones for fructose.

Onions, potatoes, corn, and carrots have a large sugar content that can be caramelised. In the case of vegetables that have only a little sugar, such as the members of the cabbage family, it helps to add a bit of fructose.

Caramelised onions.

Green cuisine comes in many colours

Maillard reactions

Maillard reactions is the umbrella term used to describe a number of chemical reactions that are able to create hundreds of different chemical compounds from sugars and proteins (amino acids) present in the ingredients. The resulting compounds are brown and are valued not only for adding colour but are also a source of pleasant tastes and aromas. Because the compounds contain nitrogen and sulfur atoms from the amino acids in the proteins, their tastes are fuller and more complex than those that are produced when sugar on its own is caramelised. Maillard compounds are described as being floral, earthy, meaty, chocolaty, coffee-like, and vegetable-like. They are characteristic of foods that many people love, such as coffee, dark beer, crackling on a pork roast, chocolate, or the crust on baked bread.

The reactions take place only if sufficient quantities of specific types of sugars and amino acids are available. Sugars such as fructose (fruit sugar), glucose, and ribose (found in nucleic acids) work well, whereas sucrose (ordinary table sugar) and maltose are less effective. Among the amino acids, lysine and cysteine very easily enter Maillard reactions. Sometimes an ingredient, for example a vegetable, has a sufficient supply of both the right sugars and amino acids. At other times, it is necessary to add a little of one or the other in order to produce the reactions.

Maillard reactions are temperature sensitive and normally take place at temperatures in the range of 100–140°C (212–285°F). Under certain conditions, however, they can also occur at somewhat lower temperatures, although this will take much longer, up to weeks and months. This is true for the aeging of certain beers and wines, as well as the production of soy sauce, balsamic vinegar, and black garlic.

At the other extreme, undesirable and potentially harmful substances, such as acrylamides, can result from Maillard reactions that take place at too high heat, over 284°C (543°F). An example is deep-frying potatoes to make chips at more than 230°C (445°F).

In addition to the taste and aroma substances, Maillard reactions produce water. If water is already present, the reactions are slowed down and too much water also makes it difficult to surpass the temperature at which water boils. One can, however, speed up the Maillard reactions by removing the accumulated water. This is why this type of browning is best carried out in oil on a skillet, in the oven, or on a grill. Browning cannot take place in a microwave either, because it works by heating the water that is in the food and, consequently, the temperature cannot exceed 100°C (212°F).

Maillard reactions are also affected by acidity and take place more quickly at a higher pH. They can, therefore, be accelerated by adding a little baking soda, which is basic, when browning vegetables.

The composition of carrots, onions, aubergines, corn, and potatoes make them very suitable for browning. Cabbage and onions contain a great deal of cysteine and, as a result, produce especially strong-tasting Maillard compounds. Vegetables such as broccoli and cauliflower contain only a little sugar, but they can be browned with the addition of fructose.

Maillard reactions often take place under the same conditions as caramelisation. An example is browning carrots and onions with the addition of fructose and glucose at about 180°C (355°F) for an hour or so. In order to ensure that the carrots do not burn, one can shorten the time by adding a little baking soda to raise the pH level. The combined effect of the two processes releases a profusion of enticing aromas and tasty substances.

Colourful seaweeds

Even though seaweeds are traditionally divided into three main groups—green, brown, and red—one cannot always rely on the colour to tell us to which biologically determined species a particular seaweed belongs. The colour can depend on variations in the amount and type of pigments in the organism, its tissue composition, where it lives, and its age.

Like plants, all species of seaweeds carry out some form of photosynthesis and have chloroplasts that contain green chlorophyll *a*. But the green colour that would normally result can be masked by pigments that have brown, yellowish, and red hues. In addition, their colours can be faded by sunlight, so that, for example, a reddish-purple or green seaweed species can appear very pale, almost white.

The colour of green seaweeds such as the various species of sea lettuce (*Ulva* spp.) is primarily due to chlorophyll *a*. The brownish-yellow colour of brown seaweeds is attributable to an accessory pigment called fucoxanthin that absorbs energy from the sunlight and passes it on to the chloroplasts. Similarly, in the red seaweeds, for example, dulse (*Palmaria palmata*), accessory pigments called phycobilins are responsible for red, orange, and blue hues. Red seaweeds often lose a lot of their colour when they are torn away from the place where they are growing and drift around in the ocean, whereas green seaweeds fade much more slowly. This is because phycobilins are water soluble but chlorophyll *a* is not.

The shapes and patterns of seaweeds can make a contribution to the aesthetic aspect of the dining experience. The visual aspect of an arrangement of seaweeds is especially highly prized in Japanese cuisine, both in salads and fish dishes. The smaller red seaweeds, which in fact do not always appear red, are often selected because they come in a great variety of shades and shapes, not least because of their multi-branched structure. Species that have firm stems and blades have an advantage over the majority of seaweeds in that they do not collapse onto each other on the plate.

Two delicate species of red Japanese seaweeds, *funori* (*Gloiopeltis* spp.) and *ogonori* (*Gracilaria* spp.), are popular. So is *tosaka-nori* (*Meristotheca palulose*), which comes in three different colours, white (*shiro-tosaka*), green (*ao-tosaka*), and red (*aka-tosaka*), and is very much in demand. As it is both crisp and colourful, it makes a wonderful contribution to a simple leafy salad. The common and very robust Irish moss (*Chondrus crispus*) is just as colourful, but its texture can be somewhat tougher, and it is not especially tasty.

Blanching brown seaweed

When brown seaweed species, for example, *wakame* (*Undaria pinnatifida*), winged kelp (*Alaria esculenta*), serrated wrack (*Fucus serratus*), or bladderwrack (*Fucus vesiculosus*), are blanched, the brown and yellow pigments break down and the green colour of their chlorophyll *a* becomes dominant.

Tips of bladder wrack (*Fucus vesiculosus*) before and after they have been blanched.

The delicate natural colours of Irish moss (carrageen, *Chondrus crispus*).

ADDING TASTE AND TEXTURE

introducing The Flavour Accelerators

The two most basic spices, salt and pepper, are found in virtually every kitchen. And in most there is also a spice shelf or drawer with a selection of dried parts of plants, especially herbs, but also seeds and bark, that are kept within easy reach so that it requires little effort to fine-tune the taste of a dish. More correctly, one ought to say aroma instead of taste because most of the traditional spices appeal mainly to our sense of smell. There are, of course, exceptions such as pepper and chilli that feel irritating in the mouth and have a strong, burning mouthfeel.

There is also another whole range of ingredients in the kitchen that we add to dishes to make them more flavourful but that are not thought of as spices. Possibly this is because some of them are liquids or are stored in the refrigerator. Examples are vinegar, lemon juice, soy sauce, *dashi*, fish sauce, and *gastrique*, all of which first and foremost affect the taste, although some of them also contribute aroma. These and many others really prove their worth when they are used to enhance green cuisine.

Based on our experiments to prepare plant-based food that is more appealing, we have singled out a number of ingredients that are especially useful. We call this collection of seasonings *The Flavour Accelerators*, because their main job is just that—to raise the sensory experience of a dish to a whole new level.

When considering how these seasonings might be used, it helps to keep in mind that they have widely differing keeping qualities. Some of them, such as lemon juice and garlic, are easy to keep on hand, are used fresh, and need only a minimum of preparation. Others, for example, *salsa verde*, *gremolata*, bouillon, and caramelised onions, are relatively easy to make but should also be used up fairly quickly. Still others that can also be prepared at home but with a bit more effort, for instance, mayonnaise and pesto, will keep somewhat longer. Finally, many commercially prepared products, including *miso*, balsamic vinegar, and fermented sauces, have a very long shelf life.

In addition to their ability to add taste and aroma, many of *The Flavour Accelerators* can contribute texture and a contrasting mouthfeel to green cuisine. This can be done by selecting different crunchy ingredients such as toasted seeds and nuts or bread croutons, as well as crisp pickled vegetables, in particular, Japanese *tsukemono*.

The Flavour Accelerators

Detailed information about *The Flavour Accelerators* listed below can be found in a separate chapter, where they are described with regard to their taste, aroma, and texture. Recipes are also provided for some of those that are easy to make at home or that are not generally for sale in grocery stores. As many of these seasonings have excellent keeping qualities, they can be bought or made up in bigger quantities and stored in the refrigerator if necessary, so that they are on hand when needed.

 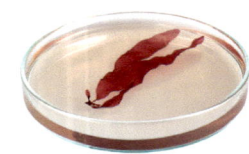 

Gallery of items representing *The Flavour Accelerators*.

Aïoli
Anchovies (paste, sauce)
Bacon
Balsamic vinegar
Black garlic
Blue-veined cheeses
Botargo
Bouillon
Bread
Capers
Cephalopods
Charred onions
Chilli peppers
Citrus fruits
Dashi
Fish flakes
Fish sauce
Fruit vinegar marinade
Furikake and *yukari*
Garlic
Garum
Gastriques
Ginger
Goma-shio
Gremolata
Ham (air-dried and cured)
Hemp seed crunch
Hoisin sauce
Katsuobushi
Ketchup
Koji
Lees
Lime-soy-fish sauce dressing
Marinated mushrooms
Marmite
Mayonnaise
Mirin
Miso
Mushroom powder
Mustard
Nutritional yeast
Oyster sauce
Panko
Parmesan cheese
Pepper
Pesto
Pickles
Ponzu and *sanbaizu*
Potato cooking water
Remoulade
Roasted nuts and seeds
Romesco sauce
Rouille
Sake
Salsa verde
Sansho
'Seaweed liquorice'
Seaweed powder and granulate
Shellfish powder
Shichimi
Soy sauce
Tahini
Tamarind sauce
Tomatoes (sun-dried and in paste)
Tsukemono
Verjus
Vinaigrette
Vincotto
Vinegar
Worcestershire sauce
Yuzu
Za'atar

Adding taste and texture

How to meet the taste challenge head-on

While most people acknowledge that it is desirable to eat plenty of vegetables and fruits, these are not always prepared in ways that spark a great deal of enthusiasm. So, one needs to start by figuring out how to ensure that these essential components of green cuisine can measure up to the meat dishes that have traditionally been considered more worthy of our attention.

We, the authors, were already familiar with a whole range of flavour accelerators that are commonly used in Western cuisine, even if they are not necessarily thought of as fulfilling that specific function. But we came to realise that there was a way to expand on the concept by incorporating ingredients that are most often associated with Asian food, only a few of which are widely used in other cuisines. In the course of our earlier work on such topics as sushi, seaweeds, and Japanese pickling, we were greatly inspired by Japanese temple food, known as 'the enlightened kitchen' (shōjin ryōri), which was described in an earlier chapter. The Buddhist monks, who are vegans, discovered early on how to make delicious food by using the soup stock dashi and a number of fermented products such as soy sauce, miso, mirin, and sake.

While we have built on some of these principles, our overall approach is flexitarian and neither vegan nor limited to a particular style of cooking. The recipes make use of a full range of flavour accelerators, including products such as fish sauce, aged cheese, air-dried ham or fish, as well as dairy and egg products. The main criterion is that these seasonings can, even in small quantities, contribute umami and koku.

Caesar salad is a perfect illustration of how this works. It consists mainly of pieces of romaine, which are pleasingly crisp but have little taste on their own. This is where a good dressing and a few trimmings such as Parmesan cheese, hard-boiled eggs, anchovies, ripe tomatoes, bacon bits, or Worcestershire sauce come to the rescue. The end result is delicious salad, as popular today as when it was first created a century ago.

Umami can have the effect of amplifying salty and sweet tastes, so that one can reduce the amount of salt and sugar in a dish, while at the same time counteracting bitterness. And koku tends to intensify all tastes, especially sweetness, saltiness, and umami, leading to a concentrated, pleasant mouthfulness and a lingering, composite aftertaste.

Creating these sensory impressions and many other cooking tricks, including the use of fermented and pickled products, take advantage of the way in which *The Flavour Accelerators* can transform plant-based cuisine. Most of them are featured in the recipes in the book.

Seven is a lucky number

In Japan the number seven, shichi, is regarded as a lucky number, not least because of the influence of Buddhism, which introduced the idea of seven reincarnations. And for the first seven days of the Lunar New Year, people show reverence toward the Seven Lucky Gods, Shichifukujin. The positive connotations of the number have also found their way into Japanese gastronomy. For example, it is said that there are seven essential ingredients, all of them in fact flavour accelerators—soy sauce, miso, konbu, katsuobushi, sake, mirin, and rice vinegar—that can be mixed to create a whole range of other tastes. In addition, there is a special spice mixture, shichimi, meaning seven types of taste, that is typically made from sansho pepper, sesame seeds, red chilli, dried ginger, aonori (a green seaweed similar to sea lettuce), yuzu peel, and hemp seeds. There are local variations in the composition of shichimi, but it always has seven ingredients.

'Greens to go'

The idea behind 'greens to go' is to have on hand ingredients that can be used quickly and easily to create a variety of salads with different tastes and textures. All kinds of firm vegetables are suitable, which helps to reduce food waste.

It is best to prepare a 'greens to go' mix from one to three types of cabbage, carrots, bell peppers, radishes, or other vegetables that are in season, and leftover bits and pieces. The ingredients should be chopped up or grated coarsely and then stored loosely in a sealed bag in the refrigerator for up to four days.

The 'greens to go' can quickly be tossed with some leafy salad greens and possibly tomatoes and cucumbers and seasoned with a dressing inspired by *The Flavour Accelerators* according to one's own preferences or to complement the rest of the meal. To add interest and texture, top the salad with garnishes such as toasted nuts, croutons, seeds, and a little grated cheese.

'Greens to go'—a bagful of mixed, chopped vegetable pieces that can be kept on hand to add texture to a salad.

Texture needs special care

When it comes to changing the texture of liquid or semi-solid food ingredients, for example, dairy products, baked goods, pasta, or sauces, one can make use of a whole range of effective thickeners, stabilisers, and emulsifiers that can change viscosity and create gels and other solid or semi-solid dishes. Many of these are sourced from seaweeds and plants—alginate, agar, carrageenan, carob powder, guar gum, gum arabic, pectin, and starch.

In the case of plant-based ingredients, however, one is usually dealing with very or fairly solid vegetables and fruits. Here the goals might be to preserve desirable textures (crispness and juiciness), to soften the fibres (heat treatment), to reduce water content (dehydrating), or to alter them altogether (fermenting, marinating, or using salt, acid, sugar, enzymes, or microorganisms). Many of the recipes in this book include concrete examples of how to do so for a broad range of raw ingredients. But in order to understand how texture is affected by these various treatments, it is useful to have some basic background knowledge about plant cells and how they make up plant tissue.

The structure and function of plant cells

The mouthfeel of vegetables and fruits can best be understood by taking into account their composition, starting at the cellular level. Their special texture when raw is due to a combination of the inner structure of the cells and of the cell walls, the way in which they are bound to each other, and their moisture content (water or juice). How these elements interrelate helps to determine how texture is affected when one prepares the vegetables and fruits with the help of salt, sugar, acids, enzymes, or microorganisms or subjects them to heating or cooling.

Dietary fibres

In contrast to animal cells, plant cells are protected by a cell wall composed of two types of carbohydrate, namely cellulose and hemicellulose, which are not water soluble. This creates the stiffness and support that enables a collection of cells to form a stable structure such as an upright stem or a broad,

Full-grown asparagus after the harvesting season.

opened leaf. As humans do not have the enzymes required to break down and digest cellulose and hemicellulose they are classified as insoluble dietary fibres. The crispness of a vegetable or fruit depends to a certain extent on how strongly the cell walls are held together by these fibres.

In between the cells there are other dietary fibres, as well as water and minerals. One of these is another carbohydrate, pectin, which typically makes up 15–20 per cent of the fibre content in vegetables. These fibres are, in a sense, the glue that holds the plants together by binding themselves to the cell walls and to each other in all directions. Pectin is water soluble. When the pectin molecules are dissolved in water they become negatively charged and repel each other. In order to bind the pectin molecules tightly to each other and form a firmer structure together with the water, that is, create what we call a gel, it is necessary to counteract this repulsion. One way to do so is to add sugar (sucrose), which binds water and, as a consequence, forces the pectin molecules to link together a little more strongly. This method is used, for example, to make a jelly with fruit juice. Another possibility is to add acid, which reduces the extent of the electrical repulsion. Still another option is to add sea salt. It contains positively charged, divalent calcium ions and magnesium ions and, in this way, brings the negatively charged pectin molecules together. This can also be achieved by adding a pure calcium salt, such as calcium citrate. Which of these possibilities results in the strongest gel depends on the type of pectin in question.

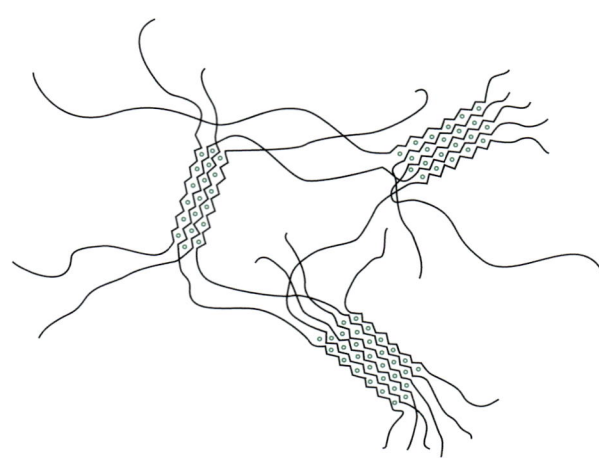

Schematic illustration of the way in which long pectin molecules are cross-linked by calcium ions (green circles) in so-called 'egg carton structures.'

The essential energy depot for vegetables is found inside the cells in the form of starches, which are complex carbohydrates made up of a large number of glucose units. In general, vegetables contain only a little fat. When they are heated and cooked their starch gelatinises, the cell walls become less stiff, and the plant tissue develops a texture that is soft and easier to chew. The cellulose is, however, not broken down and, therefore, cooked vegetables still retain a lot of their insoluble dietary fibre.

Water content and crispness

With regard to crispness, the most important factor is the presence in each cell of an organelle called a vacuole, which is enclosed in its own membrane and can occupy up to 80 per cent of the volume of the cell. Even though the name of the organelle could lead one

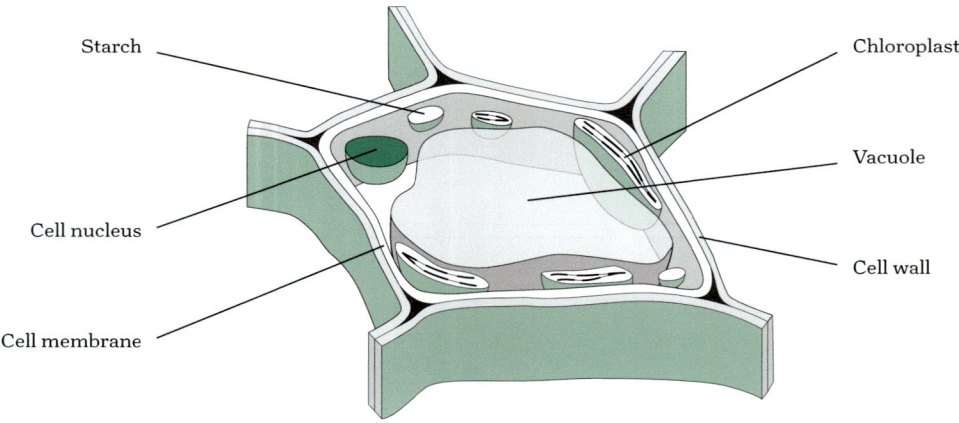

Schematic illustration of a plant cell.

to believe that it is an empty shell, this is far from the case. The cell of a juicy fresh vegetable can consist of as much as 90 per cent water, most of which is in the vacuole. Although it is primarily a water depot, it also carries out other tasks such as storing nutrients and breaking down waste products.

The fluid content of the vacuole is controlled by what is known as osmotic pressure, which is formed between the cell membrane that lies under the cell wall and the inner membrane that surrounds the vacuole. Osmotic pressure is a fundamental physical-chemical effect that can arise over a membrane that is permeable only to small molecules, such as those of water, and impermeable to large molecules, such as those of salt, sugar, and proteins. The side of the membrane on which most of the small molecules are located has a tendency to draw water through the membrane, giving rise to a difference in pressure. This causes the fluid content of the vacuole to press the cell membrane against the cell wall, resulting in what is called turgor, that is, the normal rigid state of the cell. Turgor is greatest when the vacuoles are brimful.

Lack of water will decrease the osmotic pressure in a vegetable. If a living root vegetable does not have a sufficient supply of water, for example, during a period of drought, its turgidity will be reduced and it will feel flaccid, a little tough, and less juicy. Similarly, if a harvested one has been dried or marinated in salt or sugar, it will become limp. It is possible, to a certain degree, to reverse this process by watering the soil around a vegetable that is still in the ground or immersing a harvested one in ice-cold water. If, in the meantime, the cell walls have been damaged by being heated or frozen, the cells can no longer maintain osmotic pressure and the liquid that seeps out cannot be drawn in again.

Crispness can be defined as the resistance of a vegetable to the actions of the teeth, which, if the bite is sufficiently strong, are able to fracture the cells causing the liquid in the vacuoles to spurt or seep out. The resulting mouthfeel determines whether or not we consider the plant to be crisp and juicy. If fracturing also creates a sound, which might be transmitted through the skull bones, we characterise the taste sensation as crunchy.

As a vegetable or fruit ripens, its pectin content might also be affected by the action of an enzyme called pectinase, which loosens the bonds between the cells. As a result, it may feel mealy and dry, even though its water content might be unchanged.

Restoring crispness

The fastest way to refresh vegetables, such as carrots, green asparagus, and *daikon*, that have become a little soft and limp due to water loss is to peel and julienne them and then plunge them into ice-cold water. After a short time, the cells of the vegetables will have drawn in some of the water, and they will be firm and turgid. Their ends often curl up rather decoratively.

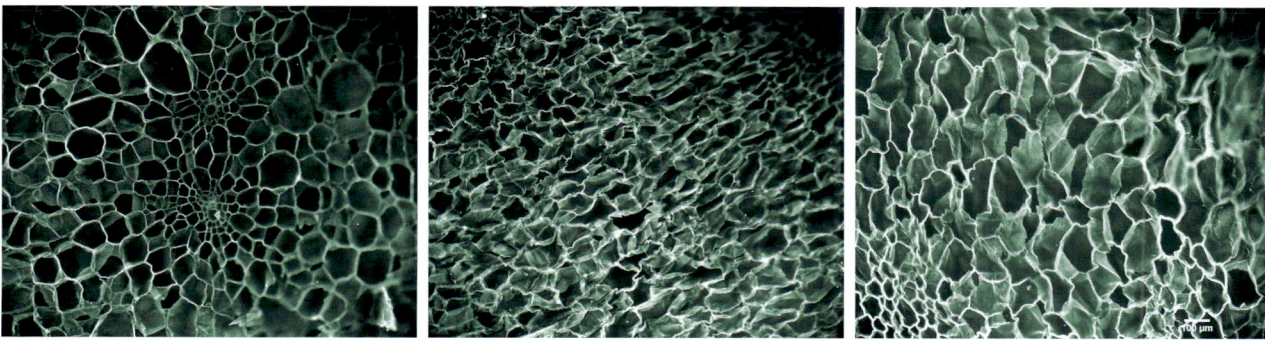

The cell structure of a radish: fresh (left), dried (middle), and rehydrated (right). Drying causes the cells to shrink and the network of the otherwise stiff cell walls becomes crumpled and more flexible. When the cells are rehydrated, they once again absorb water, but the cell wall network remains irregular and flexible. As a result, the radish is still soft, but it also feels crunchy. The actual pictures correspond to 2.0 mm × 1.6 mm.

The effect of heat on texture

Heat is very often used to alter the texture of plants, especially vegetables. This is where our knowledge about pectin and hemicellulose and their ability to bind to each other can be put to good use.

When vegetables are heated to temperatures above 60°C (140°F), their cell membranes are damaged, allowing the liquid inside the cells to seep out, leaving them limp. At higher temperatures and at the boiling point of water the links between the cellulose fibres are weakened because the molecules that bind them together, such as pectin, break down and dissolve in the water. The cellulose itself remains intact but the plant tissue has been softened to a considerable extent.

The ways in which pectin and hemicellulose interact can be used to advantage to prepare vegetables to obtain the desired texture. In this respect, the pH factor and the presence of certain ions are especially important. For example, acid from lemons or vinegar strengthen the links between the fibres and make the texture firmer, while bases such as baking soda have the opposite effect as they weaken the links and soften the vegetables. Plant cells also contain their own acid, which can seep out and change the pH level of the cooking water so that their texture becomes harder. One can minimise this effect by boiling the vegetables in a large quantity of water and preferably for a short period of time to preserve their colours. Divalent ions such as those from calcium and magnesium strengthen the links, while monovalent ions, for example, from the sodium in table salt, have the opposite effect. Using sea salt and hard water is another way to prevent the vegetables from becoming too soft.

Another approach is to cook firm vegetables such as carrots, beets, potatoes, or white beans at first at a lower temperature before bringing them to a boil, which sometimes results in a more robust texture. The same is true for fruits such as apples and tomatoes. This is because these raw ingredients contain certain enzymes that are activated at about 50–70°C (120–160°F). The activity of the enzymes makes it easier for the pectin molecules that bind the cell walls together to form a network with the help of the calcium ions that are seeping out of the cells in this temperature range as their cell membranes break down.

As vegetables are cooked, over time they become softer and lose their crispness, resulting in a mealy mouthfeel or breaking them into bits, even to the point of resembling a paste or a purée. For example, the texture of cooked white beans depends to a great extent on the cooking time and on the ions in the water. In order to prevent the beans from turning to mush, one can add a little calcium chloride to the water. The suggested amount is ½–1 g (⅛–¼ tsp) for 100 g (½ c) of water. The calcium ions strengthen the bonds between pectin molecules, which helps to keep their surface texture firm while the inside becomes softer. It can, however, be difficult to cook the beans until they are tender in areas where the water is hard and has a high concentration of calcium ions. In this case, adding a little baking soda to the water decreases the time required to make them tender.

Why do some vegetables become woody?

As a plant grows and by the time it has reached its maximum size, a substance called lignin starts to be deposited in the cell walls. It reinforces the cell walls, providing the structural strength that is characteristic of trees. Vegetables that have a great deal of lignin are sometimes described as having a 'woody' mouthfeel. Some vegetables continue to accumulate lignin after they have been harvested, with asparagus being an especially good example of this effect.

Asparagus—handle and store carefully
Asparagus are perennials that have an underground root system with fleshy storage roots attached to a horizontal rhizome and thin branched feeder roots that absorb water and nutrients. Every spring, numerous vertical shoots start to grow from the roots. As long as these shoots are under the soil, they are pale, either white or mauve. Once they are exposed to sunlight, they start to produce chlorophyll and turn green. There are also special, less common, cultivars that have a purple tinge. The stalks, usually called spears, have very small scale-like leaves that cover immature clusters of branches that can later develop fully on older plants. The spears are encased in a water-shedding skin that can be tough, especially near the bottom end. Green asparagus do not have much nutritive value with only about 2–4 per cent sugar content, although that of purple ones is somewhat higher.

The colour of purple cultivars and sometimes the tips of green varieties is attributable to the presence of the pigment cyanidin, also found in vegetables such as red cabbage and purple kohlrabi. As cyanidin is water soluble, the colour will fade somewhat when the spears are cooked.

Asparagus are eaten primarily for their taste and texture. Because the spears are designed to be stiff and to stand erect in the sunlight, they have an abundance of fibre. In addition, there are hollow vascular channels lengthwise in the spears. The fibre is mostly cellulose, which is indigestible, insoluble, and has no calories. Like starch, cellulose is made up of glucose molecules, but in this case, they are bound tightly together. As a result, the glucose in the fibres does not gelatinise when cooked in water, even though the asparagus spears do become softer.

The stems of asparagus continue to be biologically active after they have been cut away from the roots. Then enzymatic action quickly converts the very small amounts of sugar in them into lignin, the same material that reinforces the cell walls of plants such as trees. This is why asparagus, especially the white ones, and other vegetables such as broccoli become 'woody' if they are left lying around for any length of time. This process sets in as little as 12 to 24 hours after they are harvested.

When the sugar is converted to lignin the asparagus will taste less sweet. Lignin formation takes place more quickly in a warm, well-lit environment. It can be slowed down by keeping the cut asparagus in darkness at a temperature of 2–10°C (35–50°F) with a humidity level of 95–100 per cent. Once it has formed, lignin cannot be broken down so there is no solution other than cutting it away or living with having to chew the tough parts more thoroughly. But the asparagus stems can be partly peeled to remove some of the fibre. The peels can be set aside along with the less desirable woody bottom ends and cooked to make a soup stock with a delicate asparagus taste.

White asparagus are a seasonal delicacy
White asparagus are actually not a separate variety but are green asparagus that have been cultivated using a different technique. As the plants begin to grow, they are buried under a thick mound of dirt or covered with black plastic to prevent them from being exposed to sunlight. In the absence of light, the stems cannot produce chlorophyll and are colourless. Needless to say, this process, known as etiolation, is labour intensive and is reflected in the price. White asparagus are in season for only a short time and are wildly popular in Germany, France, and the Netherlands. They are considered to have the quintessential asparagus taste—mild and delicate, with only a little bitterness. On the

Green and white asparagus.

other hand, they tend to be thicker, have an outer skin that can be quite stringy and a base that can be too woody to chew. Normally the bottom two-thirds of the stems are peeled, and the end part is discarded before cooking.

After the stems have been peeled, the peels should be spread on top of them until they are to be prepared to help them keep moist and hence crisp and juicy. The turgor of harvested asparagus can be maintained by standing the spears in a 5–10 per cent sugar solution. It is important to cook them for as short a time as possible, in fact, practically not at all. If in doubt, cut off a little piece to test for doneness. And if they are to be served cold, stop them from cooking any further by plunging them into ice-cold water.

Another approach is to immerse the peels in lightly salted, boiling water and then turn off the heat. After 10 minutes, remove the peels with a slotted spoon and discard them. Reheat the water and use it to cook the asparagus. If the goal is to maximise the asparagus taste and aroma of the cooking water for possible use in a vegetable dish or with chicken meat, the asparagus can be boiled for 40–50 minutes, leaving the asparagus limp and soft.

Because white asparagus are generally thicker than the green and purple varieties, they lend themselves to roasting or grilling. This leaves them slightly charred, sweeter, and full of umami.

Why do some vegetables become woody?

Tsukemono—creating umami and interesting textures the Japanese way

Tsukemono, pickles that are made with a wide range of vegetables and a few fruits, feature prominently in Japanese cuisine. The term, which is pronounced like 'tskay-moh-noh,' literally means 'something that has been steeped or marinated,' but somewhat confusingly refers both to the techniques used to prepare them and to the finished products themselves.

The idea embodied in *tsukemono* extends to encompass a general concept of how the raw ingredients can be prepared so that they take on many desirable qualities. They should stimulate the appetite, be palatable and flavourful, have an interesting mouthfeel, give off enticing aromas, and keep well. As a bonus, their nutritional value and potential healthful effects are enhanced.

This culinary art is rooted in age-old Japanese preservation techniques that do much more than simply preserve the vegetables and fruits. It relies on a variety of conservation methods, including dehydrating, marinating in salt, vinegar, sugar, and alcohol, fermenting in soy sauce, *miso*, rice bran, sake lees, or *koji*, and aeging. The techniques used to prepare *tsukemono* resemble other well-known pickling and marinating methods, as well as fermentation caused by naturally occurring lactic acid bacteria that are usually found on the surface of plants.

The resulting *tsukemono* are especially characterised by umami taste and their fantastic crispy and crunchy textures. Cooked white rice enjoys a prominent position in the diet in this part of the world. But as it is bland and has few nutrients apart from carbohydrates, even a few pieces of *tsukemono* could be an important adjunct to a simple meal. Some Japanese chefs say that a common characteristic of most kinds of *tsukemono* is that their taste is one that lingers.

The preparation of *tsukemono* reflects the necessity in earlier times to preserve foodstuffs so that they could keep through the changing seasons. With the advent of refrigerators, freezers, and long-distance transport of fresh ingredients this has become much less important. Whereas the *tsukemono* of an earlier age could keep for a very long time, many of those prepared now have a much shorter 'best before' date because they contain less salt.

Tsukemono can have the characteristic mouthfeel of freshly harvested vegetables and be just as appealing. In the course of the changing seasons, especially in winter, they can evoke a sense of summer. Others are distinguished by tastes that are due to aeging or fermenting over a long period of time and they can often be much crisper that the fresh ingredients from which they were made.

In Japan *tsukemono* are prepared from virtually all vegetables. It is said that there are 4,000 different varieties and more than a hundred different techniques for making them. *Tsukemono* can be divided into two main types, which are differentiated by whether they can be made quickly and keep for only a short period of time or whether their preparation is more involved and drawn out. *Asa-zuke* are an example of the former. The vegetables are simply marinated in a light brine for a few hours, overnight, or a few days and will keep for up to a couple of weeks. Preservation of the latter, called *furu-zuke*, is a complex process that requires time and attention and eventually aeging over a period of many months. But these will last for up to several years. The two types of *tsukemono* have a very different taste profile.

Different kinds of *tsukemono* at a Japanese market.

In the case of most vegetables, by far the most interesting outcomes with regard to mouthfeel are achieved by first dehydrating the ingredients so that their water content is greatly reduced before they are marinated or placed in a fermentation medium. This can be achieved by curing them with salt or air drying, with the latter often leading to a better outcome.

The different ways of preparing *tsukemono* can be broadly divided into ten major categories, which are sometimes combined:

1. *Shio-zuke* (marinating in salt)
2. *Su-zuke* (marinating in wine vinegar)
3. *Sato-zuke* (marinating in a sugar solution)
4. *Amazu-zuke* (marinating in a vinegar and sugar solution)
5. *Shōyu-zuke* (marinating in soy sauce)
6. *Karashi-zuke* (marinating in mustard)
7. *Nuka-zuke* (lactic acid fermentation in bran)
8. *Kasu-zuke* (marinating in lees)
9. *Miso-zuke* (marinating in *miso*)
10. *Koji-zuke* (fermenting in *koji*)

This brings us back to the importance of the number five. Eating *tsukemono* appeals to all five of our senses—the taste on the tongue, the aroma in the nose, tactile impressions of its mouthfeel, crunchy sounds for the ears, and a symphony of colours, patterns, and shapes for the eyes. This is precisely the secret reason why *tsukemono* is such an effective tool for enhancing the overall sensory impression of vegetables prepared this way.

It is said that the contents of a meal prepared in the style of vegetarian Japanese temple cuisine, *shōjin ryōri*, should do more than simply appeal to the five senses, as listed above. The raw ingredients must also incorporate the five basic tastes, be of five distinct colours, namely, red, green, yellow, white, and black (or dark), and be prepared in five different ways. *Tsukemono*, all on their own, can satisfy these three criteria, for example, with red *beni-shoga* (pickled ginger), green *kyuri asa-zuke* (cucumbers), yellow *daikon nuka-zuke* (Chinese radishes), white *senmai-zuke* (turnip), and dark brown *uri nare-zuke* (pickling melon). If these are cut up into interesting shapes or patterns, a small artful arrangement of these pickles can not only attract attention on their own but can also make just about any vegetable dish appear more appetising and appealing.

A few sample recipes of different types of *tsukemono*, ranging from the quick and easy to those that require long-term aeging, are found at the very beginning of the recipe section. They are there to encourage the reader to dive, without further ado, straight into green cuisine. They demonstrate how simple it can be to prepare vegetables that are rich in umami and that can be eaten as stand-alone snacks or used to complement the other dishes that make up a meal.

Selection of different kinds of *tsukemono*.

Adding taste and texture

Koji has an almost magical effect on vegetables

Koji, a fermentation culture made with a filamentous fungus, has a history that goes back several millennia. Archaeologists have found traces of it on Chinese pottery that is about 9,000 years old. The first written reference to the use of the *koji* mould in food preparation can be traced back to *The Rites of the Zhou Dynasty* in 300 BCE. Over the next centuries it evolved to become a vital ingredient in Far Eastern cuisine for the production of soybean paste, soy sauce, *miso*, and sake, four ingredients that are now widely accepted as versatile seasoning agents.

In Japan, *koji* is also especially prized for its ability to transform and preserve vegetables into a type of pickle. More recently, *koji* has been introduced into modern Western cuisine as a stand-alone ingredient, where the fungus culture is used to ferment fish, meat, vegetables, beans, grains, and even insects.

A *koji* culture is made with a microscopic fungus called *Aspergillus oryzae*. As part of a fermentation process its enzymes are able to break down plant macromolecules, that is, carbohydrates, proteins, and nucleic acids, to form sugars, amino acids, small peptides, and nucleotides. These contribute sweetness and umami, the two tastes that are lacking in almost all vegetables, and as a bonus, reduce bitterness.

In principle, it is possible to prepare a *koji* culture using cooked soybeans, rice, barley, or another grain. The resulting mash is then seeded with the spores from *Aspergillus oryzae*, which are typically found on the heads of rice plants and in the air surrounding organisms that are undergoing fermentation. As there can, however, be many different variants of the fungus depending on the location and the environment, it can be difficult to obtain a uniform outcome. Because it is important to start with a pure culture, *Aspergillus oryzae* is grown in Japan on dried rice under carefully controlled conditions in seven certified laboratories. The freeze-dried spores needed to initiate the culture are mixed with rice flour and sold in packages as *koji-kin*.

Fortunately, there is a much less complicated option, which is to use a special product that contains sea salt and the active enzymes only, rather than the living fungus. It is sold as a paste, *shio-koji*, or as a filtered liquid, *ikitai shio-koji*, has a long shelf-life, and is easy to use.

Koji has an almost miraculous ability to bring out the best in all kinds of vegetables. Their texture becomes a little softer, although they retain a great deal of their crispness, depending on how long they are in contact with the culture. The slightly yeasty aroma and taste impart a full, intense, and round flavour experience with a considerable umami component. Members of the cabbage family, for example, broccoli, become much more appealing to those people, often children, who are inclined to reject them out of hand or find them too bitter.

Here is an extremely simple example of how to prepare vegetables using *shio-koji*. Cut an assortment of vegetables into small, no more than bite-sized pieces, and place them in a large plastic bag. Add *shio-koji*, inflate the bag, seal it tightly, and then shake it vigorously until the pieces are well-coated with the *shio-koji*. Place the bag in the refrigerator and shake it from time to time. To speed up the fermentation process, leave the bag at room temperature for a couple of hours before putting it in the refrigerator. After only two hours or so, the vegetables will already taste sweeter, milder, and less bitter, and the salt will have started to draw a little liquid out of the vegetables. They will be ready to eat in about two days and will keep for a week. A recipe for broccolini spears prepared this way is included in the recipe section.

Because the whole process takes place at low temperature in the refrigerator, the vitamin content of the vegetables is preserved. And as there is no need for sugar or fats, no calories are added either. Overall, *koji* makes the vegetable pieces more palatable, brings out more complex tastes, and makes them more readily digestible.

THE NUTRITIONAL UNDERPINNINGS OF GREEN CUISINE

choosing healthy options

It goes without saying that food does more than just keep us alive and that what we eat is intimately linked to our physical well-being. But it is generally acknowledged that its value also lies in being a source of enjoyment and that it has an unmistakable influence on our mental health. The saying "let thy food be thy medicine and thy medicine be thy food," attributed to the renowned Greek physician Hippocrates, has as much relevance today as it had in ancient times and should act as a catalyst for further reflection. Thanks to our ever-increasing understanding of the chemical make-up of food ingredients, we have also gained a better insight into precisely how individual components in the food interact with our body and our psyche, even though these relationships are very complex and still not fully understood.

Notable findings that have contributed to our current body of knowledge include a better concept of the importance of macronutrients such as proteins, carbohydrates, and fats, more precise information about minerals and micronutrients and their impact on metabolism, and the discovery of vitamins and hormones. We have also learned that although certain substances are poisonous, they can, in small quantities, be medicinal and protect us from a variety of diseases.

In recent years we have also gained an insight into how our interaction with food is to a great extent determined by the myriad of microorganisms that inhabit our digestive system, especially in the intestines, often referred to collectively as the human microbiota. These microorganisms, in turn, depend on what we eat for their food, which they convert to substances that have an effect on both our physical and mental well-being.

Food and good health

One would have thought that armed with our much-expanded knowledge base about food, we would now have an easier time figuring out how to strike a balance between what we like to eat

and what is best for our health. Sadly, though, the underlying science is both complex and constantly evolving, can be difficult to communicate, and is often misinterpreted. As a result, what has arisen in this seemingly ever-changing nutritional landscape is an almost head-spinning outpouring of advice about what to eat, much of it promoted by wellness gurus who each have their own ideas about how food is related to better health. The focus on plant-based foods is part of this trend.

In order to make sense of this plethora of information, there are a number of factors that should be taken into consideration when trying to evaluate the merits of eating a plant-forward diet. First of all, the ingredients are not all equally beneficial, nor are they the simple sum of the individual nutrients of which they are composed. There is a very complex relationship between the nutrients on the one hand and the way in which each person's microbiota affects how they are absorbed and digested on the other. In addition, the food culture to which people are accustomed and their state of mind play an important role. Finally, the potential benefits of plant-based foods depend on how they are prepared and the extent to which they need to be supplemented by ingredients from other sources.

This leads us back to the point made at the beginning of the book to the effect that humans by nature are omnivores and, hence, are also carnivores.

Of course, this does not imply that it is not possible to survive on a purely vegetarian or vegan diet but is an acknowledgement of the extent to which nutrients and energy derived from heat-treated animal foods have come to play a dominant role in the evolution of *Homo sapiens*. As a general rule, plants are much poorer sources than animal products of the energy and proteins required to build up muscle tissue. Moreover, it is very difficult to replace certain vitamins, fats, and essential amino acids found in animals from purely vegetable sources.

Different dietary patterns
There are many differences between vegan diets made up of plants, mushrooms, and seaweeds, vegetarian ones that also include dairy products and eggs, and those that incorporate the consumption of land and marine animals. Some of these differences involve essential nutrients that are either absent or found only in small quantities in plant sources. Seven of these are difficult or even impossible to obtain from plants: vitamin B_{12}, vitamin D_3 (cholecalciferol), taurine, creatine, carnosine, docosahexaenoic acid (DHA, an omega-3 fatty acid), and haem iron. The human body is able to a certain, although insufficient, degree to synthesise three of these substances. Vitamin D_3 can be produced in the skin when it is exposed to sunlight. Creatine can be synthesised in small quantities in the liver,

pancreas, and kidneys. Enzymes in the body can, but not very efficiently, convert alpha-linolenic acid (another omega-3 fatty acid), which is abundant in some seeds, to docosahexaenoic acid. As a result, individuals following diets that are primarily plant-based usually take supplements to maintain their overall health.

Plants, mushrooms, and seaweeds are generally rich in antioxidants, certain vitamins, and dietary fibres. Vitamin B_{12} is found in animal-derived food and also in some algae. Marine animals, such as fish, cephalopods, and shellfish, are sources of the important super-unsaturated fatty acids, often from the algae eaten by organisms lower down in the food chain on which they prey. Plants contain unsaturated fatty acids, but not the super-unsaturated ones. In addition, the ratio between omega-3 and omega-6 unsaturated fatty acids in marine sources is more attuned to our dietary needs as determined by evolution than that found in plants.

A thorough comparison of the components of the various diets should also take into account not simply the food ingredients themselves, but also how they have been prepared. The most recent research has shown that foods that are extensively processed increase the risk of cardiovascular diseases and all types of diet-dependent mortality. One reason for this, however, could be that a significant intake of processed foods is often coupled with a diet that has fewer vegetables, less fibre, and more salt and sugar.

Finally, it is important to note that the conditions under which the plants are grown can also lead to the presence of contaminants, such as heavy metals, herbicides, insecticides, fungicides, and substances that can affect our hormonal balance. One way to avoid this problem is by choosing organically grown vegetables and fruits. In so doing, though, it is necessary to accept that there can be differences in the nutrient content, aromatic substances, and taste components between the cultivars selected for ecological agriculture and their more widely available counterparts.

Macronutrients

Proteins, fats, and carbohydrates are referred to as macronutrients. They all contribute to the energy intake that is required for our bodily functions. In addition, proteins and fats provide the basic materials that are required to build up and repair cells and organs. They are the source of amino acids and of fatty acids, respectively.

Proteins

Proteins are long-chain molecules made up of some combination of the twenty-one different amino acids found in living cells. As proteins can contain from just a few to many hundreds or even thousands of amino acids, they vary widely. The small ones are also referred to as peptides. While the human body can synthesise some amino acids, there are nine that can be obtained only from our food intake. These are known as the essential amino acids: histidine, isoleucine, leucine, lysine, methionine, phenylalanine, threonine, tryptophan, and valine. There are also six amino acids that the body can make in only limited quantities and, finally, six that are considered to be non-essential. One in the last group is glutamic acid, the salt of which, glutamate, is responsible for umami taste.

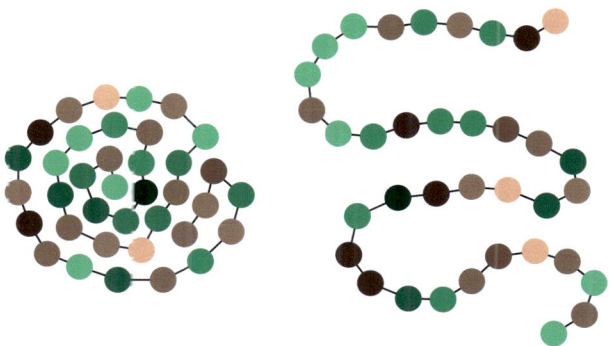

Proteins are made up of chains of different amino acids, shown here as spheres of different colours. The protein on the left is folded, while the one on the right is unfolded (denatured), for example, from having been heated.

Humans have digestive enzymes that break down the proteins into their constituent amino acids. Our cells have mechanisms and other enzymes that are then able to reassemble the amino acids into precisely the proteins we need, provided that the building blocks were present in the food that we ingest. These proteins are used both to build up structures in the cells and to carry out their active physiological and neurological functions. Many of these proteins are themselves enzymes.

The protein content of plants varies greatly, and vegetables are generally poorer sources than meat, fish, and dairy products. Some plant-based foods, however, have more than 20 per cent protein and can easily compete with the animal products. Important ones are legumes such as lentils, yellow split peas, soybeans, and chickpeas.

Edible mushrooms and seaweeds are also good sources of proteins, which in some cases can amount to 20–30 per cent of their dry weight. In this respect, however, they typically more closely resemble vegetables such as spinach and asparagus. They also contain essential amino acids, with their content varying by species, as is the case for plants. Mushrooms and seaweeds are low-calorie ingredients that do not contribute greatly to our energy needs but can add delicious tastes.

Not all proteins are equal

It is important to include all nine essential amino acids in our food intake. Such a diet is referred to as 'complete' with respect to proteins. As described above it is easiest to meet our needs from animal sources, especially fish, red meat, poultry, and dairy products. With the exception of soybeans, buckwheat, quinoa, and hemp and chia seeds, plant sources on their own lack one or more of the amino acids required for a complete diet. As a result, vegans need to ensure that their food intake includes a carefully selected variety of vegetables, legumes, and grains. For example, legumes, which are rich in lysine but have little methionine, can be combined with grains that have little lysine, but a great deal of methionine. Plant sources that have an especially high content of various proteins include beans, lentils, peas, cereals, brown rice, soybeans, hemp seeds, nuts, and certain fruits such as avocados. Even though two different protein sources, for example, from animals and plants, contain the same amino acids, they may not necessarily be equivalent in terms of nutrition. Much depends on how the amino acids in the proteins are bound together, which affects the ability of our digestive system to break down the protein molecules into usable components.

Leek flower.

Proteins and the transformation of the Japanese diet

Although the Japanese diet is often singled out as being one of the healthiest eating plans in the world, this has not always been the case. Until World War II, the average daily meals consisted of cooked white rice, a little *miso* soup, a few herbs, and pickled vegetables, with fish making only a rare appearance and meat not at all. This diet was sorely lacking in proteins and people were generally undernourished. Umami, the basic taste that is linked to deliciousness, was noticeably absent in the food eaten in an earlier time and many of the dishes that we now associate with Japanese food culture actually originated in China and Korea. How did the transformation of this meagre, subsistence diet into a tasty and healthy modern cuisine come about?

Much like other cultural transformations, a change in eating habits generally takes place over a long period of time. In the case of Japan, food historians have identified three distinct steps in this process. The first milestone is linked to the Meiji Restoration in the second half of the nineteenth century. At this time, Japan ended a long period of isolation and began to open up to the rest of the world and to move slowly from an almost exclusively agricultural society where each family grew its own food toward a capitalist industrialised one. In 1872, Emperor Meiji announced that he had started to eat meat, breaking with a thousand year-old Buddhist tradition. This paved the way for the introduction of new ingredients and people were taught to eat meat and dairy products in order to gain the physical strength needed to compete with Western countries.

The second turning point occurred in the early 1920s when the leaders of the Japanese army fighting against Russia in Siberia had to admit that their soldiers could not get by on a diet of rice, *miso* soup, and vegetables. Pork, beef, and poultry meat were included in the field rations to add proteins and fat, barley replaced part of the rice, and the vegetable content was increased. The meals were more nutritious and, being non-traditional, they did not cater to regional food preferences. The soldiers had to adapt and learn to eat this way if they wanted to survive. By the time soldiers had finished their military service, they had acquired a taste for these dishes and spread knowledge of them throughout the country when they returned home. In addition, popular publications and educational activities further disseminated the military's ideas about a healthy diet.

The final step in the transformation occurred after World War II, when the

deprivations of war time were followed by a catastrophic period of nation-wide famine. The loss of rice-producing colonies and poor harvests led to serious rice shortages. These were offset by the arrival of large quantities of wheat initially supplied as emergency humanitarian aid by the United States, which led to the widespread acceptance of dishes such as *gyoza* and *ramen* noodles. Eventually there was a dramatic shift in the sources of nutrition, substituting proteins in the form of eggs, meat, and fish for some of the carbohydrates. Taste preferences turned toward what we now regard as a traditional, healthy Japanese diet. The earlier simple rice-based daily food was to a certain extent supplanted by foreign dishes such as noodle soups, fried, simmered, and cooked vegetables, fish, and meat, all seasoned to focus on umami and served in moderate portions.

As this demonstrates, there was nothing particularly inherent in Japanese food culture that led to the healthy diet that is much admired today. Rather, it was a series of historical circumstances that shaped the population's willingness to learn to eat differently and to appreciate novel ingredients. Undoubtedly, this fusion of traditional cuisine, whether that of the ordinary people, the aristocrats, or the Zen monks, with foreign influences, was underpinned by an understanding of how to introduce umami, in particular with the soup stock, *dashi*.

It is not surprising that many Japanese are now worried that globalisation has caused the pendulum to swing too far in the direction of Western food. The incursion of a fast-food culture that is chock-full of carbohydrates, saturated fats, and a load of empty calories bodes ill for maintaining a healthy diet. Contemporary patterns of family life and the demands of the workplace are also encroaching on the time available to prepare the daily meals from scratch. Ready-made foods are an easy solution, but they also tend to have a higher fat, salt, and sugar content.

Fats

Fats are a large class of substances that the body uses to build up structures in the cells and organs. It can also store a surplus of fats from the food in special fat cells or fat depots that can later be burned up to produce energy. Many of the fats, for example, lipids and triglycerides, are made up of fatty acids. In addition, there is a particular group, called sterols, including the well-known cholesterol.

A fatty acid consists of a long chain of hydrocarbons, that is, a string of carbon atoms with hydrogen atoms along its length. Neighbouring carbon atoms are chemically bound together by either a single or a double bond. The greater the number of double bonds in the chain, the more unsaturated the fatty acid. If there are no double bonds, the fatty acid is said to be fully saturated. This is the type of fatty acid that is often found in animal products such as milk, cheese, pork, and beef. Vegetable oils are generally primarily unsaturated, as in the case of olive oil, which has only a single double bond.

Essential fats

Just as there are essential amino acids, there are essential fats that our bodies cannot synthesise and, therefore, we must obtain them from what we eat. The essential fats have two or more double bonds and make up two families, called omega-6 and omega-3 fatty acids, and the extent to which they are unsaturated is important. The two families are derived, respectively, from linoleic acid, which has two double bonds, and alpha-linolenic acid, which has three double bonds. These two essential fatty acids are the starting point for the synthesis in the omega-6 family of arachidonic acid (AA), which has four double bonds, and in the omega-3 family of eicosapentaenoic acid (EPA), which has five double bonds, and docosahexaenoic acid (DHA), which has six double bonds. AA, EPA, and DHA, referred to as super-unsaturated fatty acids, are especially important for the proper functioning of our cardiovascular and nervous systems, including the brain. They also carry out a long list of signal functions in the body, including control of inflammatory conditions.

The human body is not able to any great extent to convert fats from one family to the other. Similarly, we lack the enzymes needed to synthesise sufficient quantities of AA, EPA, and DHA from linoleic acid and alpha-linolenic acid. It is, therefore, important to obtain both types of fatty acids directly from what we eat.

Major sources of omega-3 fatty acids are saltwater fish, shellfish, seaweeds, microalgae, meat, and egg yolks. Omega-6 fatty acids are most easily obtained from green plants and their fruits, for example, sunflower oil, soybeans, and corn. Flax seeds are also an important source of alpha-linolenic acid. But research has shown that only about 5 per cent of the alpha-linolenic acid that is ingested is converted to EPA and DHA, found primarily in algae, fish and shellfish. Women seem to be better able than men to synthesise EPA and DHA. The conversion process can be further limited if there is an excess of saturated fatty acids and omega-6 fatty acids in the food. It is, therefore, much more effective to obtain the necessary omega-3 fats from fatty fish than from flax seeds.

The fat content and type of fatty acids vary greatly from one plant or fruit to another. Fruits generally are not fatty and are not good sources of unsaturated fats, with avocados and olives being notable exceptions. Dried seeds and nuts, such as chia seeds, hemp seeds, flax seeds, pumpkin seeds, soybeans, kidney beans, almonds, and walnuts are good sources of omega-3 fatty acids, specifically from alpha-linolenic acid. Herbs and plants with green leaves have reasonable amounts of omega-3 fatty acids. Plants and herbs that grow in the wild, for example, common purslane, often contain more omega-3 than ones that are cultivated.

Edible mushrooms generally have a fat content of only 2–8 per cent of their dry weight, but up to 75 per cent of these fats can be alpha-linolenic acid.

Seaweeds and algae occupy a special place in the food web because they are the original green source of the vital fatty acids that are not found in plants and mushrooms. Although their fat content is low, it is to a large extent made up of super-unsaturated fatty acids such as AA, EPA, and DHA. These bioaccumulate as they are ingested by animals and move up through the food chain.

Omega-6 and omega-3—a delicate balance
Even though the overall omega-6 and omega-3 fatty acid content of our diet is vitally important for both our physical and mental health, the ratio, omega-6/omega-3, between the two types is even more crucial.

This can possibly be explained by our evolutionary history. It is thought that early humans lived in coastal areas where there was an abundance of fish, shellfish, and seaweeds. These marine organisms contain approximately equal amounts of omega-6 and omega-3 fatty acids, typically in the range of omega-6/omega-3 being 0.3–2. In some species, there is an even greater proportion of omega-3. As our nervous system and brain have equal amounts of omega-6 and omega-3, a 1–1 ratio between the two is highly desirable in order to reduce the possibility of neurodegenerative diseases and mental disorders.

The diet in many parts of the world is now heavily weighted in favour of omega-6. For example, in Western Europe the ratio between omega-6 and omega-3 is typically 6–8, whereas in the United States it is 20 and in Mexico 25. These very high figures mirror a change in our dietary patterns to eating less marine-based food and more sourced from plants, especially inexpensive oils that have an excess of omega-6 fatty acids.

Omega-6 fatty acids are not, *per se*, harmful. But as the quantity of food that we eat is more or less constant it means that there is a relative increase in the proportion of omega-6 and decrease of omega-3. The result is to dilute the positive effect of the latter.

Seaweeds and algae are the source of super-unsaturated fats

We often say that fish are the source of the important omega-3 fats, EPA and DHA, which is true, as far as it goes. Fish and shellfish, or for that matter any other animals, are not, however, the original source of either the super-unsaturated omega-3 fatty acids (for example, EPA and DHA) or the super-unsaturated omega-6 fatty acids (such as AA). These originate in tiny microalgae, also known as phytoplankton, at the base of the food web and the large marine macroalgae, that is, seaweeds. Only these organisms have the particular enzymes that are required to synthesise super-unsaturated fatty acids. When the algae are eaten by fish and shellfish, they bioaccumulate and move up through the food chain.

Incorporating seaweeds into green cuisine is, therefore, an interesting source of polyunsaturated fatty acids, typically with an excess of omega-3 fatty acids. This is a nutritional composition that is close to the ideal balance between omega-6 and omega-3 fatty acids.

The fat content of seaweeds is quite varied. For example, dulse (*Palmaria palmata*) and *wakame* (*Undaria pinnatifida*) can be made up of 4–5 per cent fat. The polyunsaturated fats account for 30–70 per cent of the total, with omega-3 and omega-6 fatty acids making up by far the greatest proportion. Seaweeds have only very little DHA but are a source of both EPA and AA. Microalgae have all three. By way of comparison, it should be noted that these fatty acids are not found in any of the plants that form a regular part of the human diet.

Cholesterol and other higher sterols
Higher sterols are a group of waxy fat-like substances that occur naturally in animals, plants, mushrooms, and seaweeds, respectively. They help to regulate biological processes and mechanically stabilise the cell membranes.

The most commonly known sterol is cholesterol, which is found only in animals and is used to strengthen their cell walls. It is also involved in the production of hormones, vitamin D, and bile salts, but it is not an essential fat as it can be synthesised in the liver. The distribution of cholesterol and its transport in the body is controlled by certain lipoproteins. An imbalance between this transport system and the liver's capacity to synthesise and break down cholesterol can cause it to be deposited in the arteries, possibly leading to atherosclerosis, a chronic inflammatory condition. Consequently, it is still considered prudent to limit its intake to follow a healthy diet. Recent research has shown, however, that the problem is very complex. A variety of factors, among them the interaction with other fats as well as an individual's genetic inheritance, exercise regime, and mental condition also play a role.

The other higher sterols—phytosterol, ergosterol, and fucosterol—are found in plants, mushrooms, and seaweeds, respectively. As these completely lack cholesterol, they are considered a good counterbalance to foods of animal origin in which it is abundant. Brown algae, in particular, have an especially high fucosterol content. Only about 5 per cent of the sterols ingested from these green sources are absorbed in the course of digestion. But because people cannot convert one type of sterol to another, they normally contribute little to an individual's cholesterol reading. Recent research has shown that sterols from plants and seaweeds actually counteract the absorption of cholesterol by the intestines and can help to reduce the amount in the blood.

Carbohydrates

Carbohydrates are large molecules that are made up of different sugar groups (saccharides), for example, glucose (grape sugar), fructose (fruit sugar), and galactose, which are simple monosaccharides. When paired they form other sugars: among the common ones are the disaccharides sucrose (fructose + glucose, ordinary table/white sugar), maltose (glucose + glucose, malt sugar), and lactose (glucose + galactose, milk sugar).

A large number of sugar molecules can also bind together as polymers, called polysaccharides, either in the form of long chains or of branched networks. For example, individual glucose molecules join up in a chain or as a network to make, respectively, amylose and amylopectin, the two components of starch.

There are two varieties of starch: amylose (on the left), composed of a long chain of glucose molecules, and amylopectin (on the right), which is a branched network of glucose molecules.

Sugars and some polysaccharides, including starch in the case of humans and inulin in the case of plants, are important energy sources for the cells. Others, such as cellulose, lignin, chitin, glucan, and galactan provide fibre necessary to build structures and cell walls in plants, fungi, and seaweeds. In seaweeds, alginate, carrageenan, and agar among others help to bind the cells together. Some of these polysaccharides, for example, starch, inulin, alginate, carrageenan, and agar, are water soluble, while others, such as cellulose and chitin, are not.

Breaking down carbohydrates

Carbohydrates (polysaccharides) can be broken down (hydrolysed) by enzymes that simply cut them up into the single sugar groups of which they are composed. Human bodies have some of the enzymes required for this process. For example, the enzyme amylase is able to convert starch to glucose, thereby producing energy and calories. As amylase is present in our saliva, starch can taste somewhat sweet.

We are lacking the enzymes that can break down other carbohydrates, leaving them intact until they reach the large intestine where there may be microorganisms with enzymes that can do the job. As a result, no energy is released from these carbohydrates, and they are classified as dietary fibre or roughage, a topic to which we will return later. Nevertheless, it is often possible to break down the dietary fibre found in some ingredients by fermenting the foods before eating them.

The amount of energy we derive from the sugar groups and carbohydrates in our food is dependent on a whole range of factors. What is meaningful in terms of a given food's impact on our health is the extent to which it affects the blood sugar reading and the associated variations in our insulin level. This can be quantified using the glycaemic index, a scale that indicates how quickly carbohydrates are converted into glucose, classifying those where it happens slowly as 'good' and those that are converted quickly as 'bad.'

Glycaemic index
Plant-based foods are an important source of the carbohydrates in our diet. They provide about 60 per cent of our energy needs and release glucose into the blood stream. But different types of carbohydrates have varying abilities to release glucose because they are broken down at different rates in the digestive system. In order to compare different carbohydrates and the foods in which they are found, a scale has been developed to measure the relative rise in blood sugar two hours after the food has been eaten. This scale, called the glycaemic index, differentiates between the various types of carbohydrate.

The glycaemic index is not totally precise because the formation of glucose from a given food source is affected by its overall make-up, which may include fats, protein, acids, and salts, the degree of ripeness of fruits, and how the food is prepared. For example, white flour releases glucose more quickly than whole wheat flour and cooked vegetables more quickly than raw ones. Furthermore, formation of glucose varies from one individual to another. And there is not a simple relationship between the glycaemic index and the production of insulin in the bloodstream. Finally, the rate at which glucose is released is also dependent on the type of carbohydrate in the food.

Nevertheless, the glycaemic index can serve as a good guide when weighing the nutritional value of different foods. It is based on a scale of 0 to 100, where 100 is equal to pure glucose. The scale is divided into three parts: high 70–100, medium 56–69, and low 0–55. Carbohydrates with a high glycaemic index are sometimes referred to as 'white' carbohydrates.

Glycaemic index	Food sources
High (70–100)	White bread, white potatoes, polished rice, highly processed cereals
Medium (56–69)	Brown rice, unpeeled potatoes, bananas, raisins
Low (0–55)	Beans, walnuts, whole grains, most vegetables, mushrooms, seaweeds, red fruits, sesame seeds, sunflower seeds, acorns

Fennel.

The potato—a somewhat neglected vegetable

Botanically speaking, the potato is a vegetable, although it has fewer bioactive plant substances and contains more starch than other plants and root vegetables. As a result, we tend to dismiss it as not being a 'real' vegetable, much like white rice. And it does not count as part of the recommended daily intake of 500–600 grams (ca. 1 ¼ pounds) of vegetables and fruits. Nevertheless, potatoes, especially if left unpeeled, have fewer calories and more important micronutrients (iron, potassium, and zinc) than white rice and pasta and the peels are a source of dietary fibre. Yet, in spite of their many positive qualities and long history as a vital food source, potatoes have, rather unfairly, fallen out of favour in some Western countries in the past couple of decades.

The history of the potato can be traced back to the Andean regions of South America, where it has been grown for many millennia, long before the earliest archaeologically verified traces dating to 2500 BCE. The tubers, which are hardy and can be grown at many altitudes, became a staple food in these mountainous areas for both humans and their pack animals. Potatoes are often associated specifically with the Inca culture of Peru, where people planted an almost chaotic diversity of cultivars, often in the same fields. The International Potato Centre in Lima, Peru has collected and now maintains an active gene bank of almost 4,500 traditional varieties.

In the mid-sixteenth century, Spanish conquistadors learned to eat potatoes from the locals and loaded them onto their ships as basic rations for the return voyage across the Atlantic. Historical records indicate that they were planted initially on the Canary Islands for export to Antwerp and then spread slowly to other parts of Europe. The tubers were at first regarded with suspicion because they grew beneath the earth and the plants looked like other members of the nightshade family, some of which are deadly and associated with devils and witchcraft. Disdained as starchy and tasteless, they were thought to be fit for human consumption only to stave off starvation and used mostly as animal fodder. Nevertheless, as potatoes were also a staple food for Spanish armies fighting wars in Europe, they were incidentally distributed to peasants along the way. Unlike stores of grain that could be pillaged or burned, potatoes in the ground could survive the destruction and were easy to grow even in poor soil.

"You are to make the benefit of planting this crop clear to the lords and subjects and advise them to undertake the planting of potatoes this early year as a very nutritious food… Wherever there is an empty space, the potato should be cultivated, since this fruit is not only very useful to use, but also so productive that the effort put into it is very well rewarded."

Frederick the Great's Potato Order,
24th of March, 1756

A series of events in the second half of the eighteenth century finally secured the place of potatoes in the European diet. The Prussian Emperor, Frederick the Great, grasped the potential of potatoes to become a valuable alternative to bread and cereal crops and used them to feed the troops. In 1756, his official proclamation mandating the cultivation of potatoes was met with little enthusiasm. The town of Kolberg responded to the edict as follows: "The things [potatoes] have neither smell nor taste, not even the dogs will eat them, so what use are they to us?" During a subsequent time of famine, he ordered the peasants to eat them, supposedly going so far as to threaten to cut off the nose and ears of those who refused to do so, all to no avail. He then hit upon the cunning idea of growing them in fields around Berlin and posted guards to prevent theft. Sensing that they must be valuable, the hungry peasants soon started to poach them, plant them, and eat them.

At about the same time, a French pharmacist, Antoine-Augustin Parmentier, serving in the Seven

Potatoes in a variety of types and colours.

Years' War (1756–1763), had spent five years in on-and-off captivity in Prussian prisons. There he ate almost nothing but potatoes, which he credited with keeping him in good health, and became an advocate for the humble tubers for the rest of his life. In the years before the French Revolution when bread prices rose steeply, he tirelessly championed potatoes as a highly nutritious alternative. Louis XVI and his queen, Marie-Antoinette, became converts to the cause, he wearing potato flowers in his lapel and she in her hair. Given that potatoes yield three times as many calories as grain per hectare and can grow in a wider variety of soils and climates, a great proportion of the European population gradually became completely dependent of this crop.

During the seventeenth and eighteenth centuries, intermittent crop failures and hunger had been a constant threat in Europe, especially in the rural areas. Being able to plant potatoes on grain fields that were routinely left fallow every second year effectively doubled the food supply, so that famine virtually disappeared in an area that stretched from Ireland to the Ural Mountains. Having a more nutritious diet also helped to reduce the death rate from endemic diseases such as tuberculosis, measles, and dysentery, leading to a population explosion.

Nevertheless, this potato revolution that was promoted so enthusiastically as a way to feed the masses had serious implications for the future of agriculture. Although potatoes can be grown from seeds it is a tricky undertaking and, at first, they were propagated solely from cuttings of mature tubers, meaning that they were clones of the original plants. All of Europe's potatoes at the time descended from only the few tubers that had arrived with the Spaniard. The situation was exacerbated by the publication in 1840 of a treatise by an organic chemist, Justus von Liebig. In it he extolled the virtues of fertilising plants with a nitrogen source, in particular guano, the dried excrement of sea birds and bats, ironically also sourced from Peru. Their use led to increased yields and that encouraged the planting of ever larger fields with a single crop—monoculture had arrived.

It is thought that potato blight (*Phytophthora infestans*), a water mould that devastates potato crops, probably entered Europe in shipments of the fertiliser. Because growers planted only a few potato varieties, there was a loss of diversity, the plants had less natural resistance, and the diseases spread easily.

A notable illustration of the dangers of single-crop dependence is the Irish Potato Famine of 1845–1852. At the time, poor Irish people subsisted almost entirely on a diet of potatoes, mostly a single variety called Lumpers. When potato blight infestations ruined up to one-half to three-quarters of the crop seven years in succession, roughly one million Irish starved to death and at least another million emigrated.

About three decades later, it was discovered that both the mould and another potato pest, the Colorado potato beetle, could be killed by spraying a solution of copper sulfate, known as Paris Green, and lime on the plants. This development is regarded as the starting point for the modern pesticide industry. Sadly, the pests evolved in response to the use of the chemicals, touching off the search for

The potato—a somewhat neglected vegetable

newer and stronger treatments that continue to this day. It should be noted, however, that responsible growers now tend to use pesticides cautiously, not least because of the added cost. Fungicides are used to protect the plant from infection by potato blight, for which there is no cure. Fortunately, these chemicals are sprayed on the leaves and make no contact with the tubers growing underground.

Globally, potatoes are now the fifth most important food crop after sugar cane, corn, wheat, and rice. In 2020, about 360 million metric tons encompassing about a thousand different varieties were harvested, with China and India accounting for about one-third of the total. At least one billion people now depend on potatoes for some of their nourishment as consumption is increasing in the developing world, while it is tapering off in developed countries. On a per capita basis, Belarussians eat more potatoes than those of any other country, an incredible 175 kilograms (386 pounds) annually.

In the last few years, the potato's image has suffered from its being, somewhat unfairly, included in the group of carbohydrates labelled 'empty white calories,' notably polished white rice and white bread. Unlike these, potatoes are very nutritious and compare favourably with other dietary mainstays. While water makes up 70–80 per cent of their total weight, there are about 75 calories per 100 grams (3 ½ ounces), mostly in the form of starch, resulting in a high glycaemic index. But, unlike wheat, they have no fat, and their protein content is greater than that of corn. In addition, they are cholesterol- and sodium-free, and are excellent sources of Vitamins C and B_6, potassium, and important minerals. Potatoes are also a source of pigments that act as antioxidants, for example, carotenoids (lutein, zeaxanthin) in yellow-fleshed ones and anthocyanin in purple varieties.

As mentioned earlier, potatoes that have turned green contain bitter, toxic alkaloids (solanine, chaconine) and should not be eaten. In principle, it is possible to live on a diet consisting only of potatoes together with fish or dairy products.

The problem with potatoes, if indeed there is one, is that some people do not really know very much about potatoes or how to prepare them. And they are not aware that potatoes come in a much greater range of varieties than those found commonly in grocery stores. There is more to potatoes than the distinction between white, yellow, and red ones or those that are best for baking versus small nuggets for steaming.

In terms of preparation, there are a number of ways that immediately spring to mind: cooked potatoes with lots of gravy, mashed potatoes made with butter, and deep-fried chips. While delicious, these dishes may be less healthy options because of their fat content. Also, in these examples the nutritional value of the dish is diminished by peeling the tubers. Because half of the nutrients and a great deal of the dietary fibre in potatoes are found in the skin, it is a shame to peel them and discard the skins. Also, when potatoes are boiled in water, many of their umami substances seep out into the liquid. It should be kept and can be used in sauces and soups or to steam other vegetables that lack umami, thereby cutting down on fat and salt.

Flowering potato plants.

Dietary fibre

The tissues of plants, seaweeds, and edible fungi, which are built up with the help of fibres in the form of carbohydrates, cellulose, and lignin, are our only sources of what is known as dietary fibre. Humans are unable to digest these fibres because we lack the enzymes necessary to break them down into their component parts. As a result, they are not absorbed in the small intestine and contribute neither energy nor calories. Nevertheless, they are considered an essential component of a healthy diet.

Dietary fibre is classified as either soluble or insoluble. Some well-known soluble fibres are pectin and inulin from plants, and agar, carrageenan, and alginate from seaweeds. They have very good water-binding properties that are exploited to make thickening and gelation agents for use in cooking. In the digestive tract they serve to hold together the waste products before they are excreted. Insoluble fibres help to bulk up food as it moves through the stomach and intestines, which determines the rate at which it can pass through the system.

Both types of fibre can bind with minerals and metals, with both positive and negative effects. On the one hand, this can lead to unwanted loss of minerals but, on the other hand, this prevents potentially toxic materials from being absorbed, allowing them to be excreted before they can do any harm. The balance between these two outcomes depends on a large number of other factors, especially the overall composition of the food and how the bacteria in the large intestine are able to break down the dietary fibres.

A large intake of dietary fibre is advantageous as it reduces the blood sugar and fat content of the blood. It also has the effect of binding cholesterol, thus lowering the content of both free and bound cholesterol in the bloodstream.

Although dietary fibre is not a nutrient *per se*, it provides a banquet for the microorganisms in the gastrointestinal tract. As described below, there has been increasing recognition of the extent to which our overall health is tied to ensuring that these microorganisms are nourished appropriately.

As is the case with the components that make up other green cuisine ingredients, the dietary fibre content varies widely from one vegetable and fruit to another. Examples of vegetables with a high dietary fibre content are beans, beets, carrots, cabbage, and cauliflower. Similarly, kiwi, apples, pears, bananas, and strawberries are also fibre rich.

Edible mushrooms are an often-overlooked source of both soluble and insoluble dietary fibre. They contain complex carbohydrates such as chitin and glucan that can make up 30–40 per cent of the dry weight. Seaweeds generally are also excellent sources of both types, typically 25–75 per cent of the dry weight, which exceeds the proportion in foods that grow on land. Some of these, for example, galactan, xylan, alginic acid, glucan, and porphyran, are very complex carbohydrates.

Dietary fibres—the inside story
Dietary fibres have recently been recognised as having a much more important effect on our overall health than simply 'helping our digestion and keeping our intestines in working order.' Although the human digestive system cannot break down and convert them to nutrients and energy, they provide food for the myriad of microorganisms that thrive in our intestines. The total number of these microorganisms—mostly bacteria, archaea, and fungi—is estimated to be three to ten times greater than the number of cells in the human body. This collection of microorganisms that has colonised our gastrointestinal system and without which we cannot live is referred to as the human microbiota. It can be composed of as many as 40,000 different species, a diversity that seems to have a considerable impact on our health. Some of these species are harmless, while others are potentially harmful. The various microbes interact and usually compete with each other to make use of the available nutrients that keep them alive.

The different soluble and insoluble fibres are found especially in vegetables, pulses, lentils, cereals, nuts, seeds, fruit, and seaweeds. They are sometimes referred to as prebiotics, in the sense that they enrich the environment in which the microorganisms live, with a consequent positive effect on our health. Some of this is due to the enzymes used

Young red beets in the field.

by the bacteria to break down the fibres and that can create metabolites that have signalling functions in the body and that can influence processes such as cell growth and immune responses.

Our microbiota is a veritable enzyme factory that can muster about 168 different types to break down the dietary fibres. This creates energy for the growth of the microorganisms as well as for the passage of the food from the small to the large intestine. Another important function of all this activity is the release of vital minerals, such as copper, calcium, and zinc, that are otherwise bound to the dietary fibres and, consequently, might not be absorbed as they move through the digestive system. In addition, the main metabolites created in the course of the digestive process are some small, short-chain fatty acids. Their role in transmitting signals in the body has an impact on the nervous system and brain function and, consequently, on our cognitive and mental states.

This is where our food intake comes into the picture. The composition of our diet determines the make-up of the human microbiota. Significant changes to what we eat, whether due to external circumstances or ill-advised choices, have a knock-on effect on the availability of the dietary fibre for which the microorganisms are competing. This can lead to chronic illnesses such as inflammatory bowel disease, allergies, autoimmune diseases, fatigue, obesity, and cancer.

Short-lived and periodic changes in the food intake do not seem to have a long-term effect on our microbiota. But there is no doubt that the expanded consumption of industrially produced and highly processed foods with more meat and a lower fibre content have had an impact on it. As a result, some types of bacteria have been wiped out and cannot be re-established in the digestive tract.

Other factors, such as the use of antibiotics and food that is generally more germ-free, also have an impact. It should be noted, however, that these vary from one individual to another, whereas the big picture with the competition between the microorganisms and their sensitivity to our food intake do not

Dietary fibre

depend on ethnicity or the individual's metabolism.

The average Western diet, in which meat and fats are prominent, includes less dietary fibre, and that, in turn, leads to a decrease in the availability of short-chain fatty acids. This can result in conditions where infections can take hold in the intestines and their mucous membrane breaks down, increasing the potential for obesity and developing colorectal cancer.

The question then arises as to how much dietary fibre is required to support a healthy microbiota. A starting point to finding the answer would be to determine how much fibre was consumed by our ancestors over the eons when the symbiotic relationship between humans and their colonic bacteria was formed.

It is difficult to know what humans ate so long ago, but a comparison can be made with the groups of people who eat food that is not produced industrially. Their daily fibre intake is calculated to be at least 50 grams (about 2 ounces). These populations, broadly speaking, rarely suffer from inflammatory bowel diseases. Before the introduction of agriculture, the daily total was probably closer to 100 grams (3 ½ ounces). In many parts of the developed world the daily intake is now only about 15 grams (½ ounce), considerably less than the recommended minimum of 20 to 25 grams (about ⅘ ounce) for women and 30 to 38 grams (about 1 ¼ ounces) for men.

Nevertheless, it is not a simple matter to change one's diet to include more fibre because our modern microbiota is not accustomed to handling it in larger quantities. Doing so can lead to unwanted consequences such as bloating, constipation, stomach aches, or diarrhoea. It would seem, therefore, that the best approach to meeting our dietary fibre needs from plant-based foods and seaweeds is to move toward the ideal slowly, incorporating a variety of sources and adjusting them as necessary in order to deal with any problems that might occur.

The sushi factor

A group of French researchers looking at the bacterial breakdown of a particular polysaccharide, porphyran, which is commonly found in laver (*Porphyra/Pyropia* spp.), a red seaweed that is used to make *nori* sheets for sushi, discovered some surprising results. They found that the microbiota varies not only from one individual to another but also that the intestinal bacteria of an entire population was quite distinct. Porphyran is normally considered to be a dietary fibre that humans cannot digest. Contrary to this, the research showed that there was a group of Japanese who had intestinal microorganisms that produced enzymes that could break down porphyran, whereas a comparable group of Americans could not. A possible explanation was that the centuries-long tradition of eating *nori* and sushi had at some point caused the enzyme-coding gene to pass from a particular marine bacterium into a microbe found in the intestines of the Japanese. This effect was dubbed 'the sushi factor.' It is thought that something similar might have happened with other complex polysaccharides, such as agar. This story is an example of how the diversity of our intestinal flora can be shaped over time by what we eat.

Nori sheets made from the red seaweed laver (*Porphyra/Pyropia* spp.)

Antioxidants

Oxidation is a natural consequence of living in an oxygen-rich atmosphere. A variety of oxidising metals are also found in the proteins and enzymes of all living organisms. Mitochondria, the organelles that produce the energy needed by cells, create an oxidated environment that can have a harmful effect on the fats in cells and their DNA. Unsaturated fatty acids are particularly susceptible to this process. The harms accumulate in the course of the organism's life span and are also amplified by exposure to ultraviolet rays from sunlight. Plants are especially vulnerable to oxidation because they are veritable energy factories that depend on photosynthesis. In order to counteract these harmful effects, plants and seaweeds have developed a whole arsenal of substances, namely antioxidants, that can mitigate oxidation. The health-promoting effects of plant-based foods are partly attributable to these antioxidants.

This impressive assemblage of antioxidants is primarily composed of carotenoids, chlorophyll, phenols, and vitamins A, C, and E. Even though the specific content of the various antioxidants can vary widely, vegetables such as carrots, onions, garlic, asparagus, beets, spinach, and green beans are especially good sources. Likewise, cherries, blueberries, and strawberries are at the top of the list of fruits and berries.

Glutathione is a particularly important antioxidant, sometimes referred to as a 'super antioxidant.' Most people have probably never heard of this small protein made up of three amino acids—glycine, cysteine, and glutamate—even though it is one of the most ubiquitous molecules in the body. It eliminates toxins from the intestines, kidneys, lungs, and liver by making them water-soluble so that they are excreted in urine and sweat. Glutathione is synthesised in the body from the amino acid building blocks but is also found in some sulfur-rich plants, such as spinach, asparagus, avocado, and papaya, among others. A few years ago, researchers discovered that many fungi, in particular porcini mushrooms (*Boletus edulis*), are without doubt the best dietary sources of glutathione.

Seaweeds are also a veritable gold mine of antioxidants, for example, carotenoids, tocopherol, phycobilin, and fucoxanthin, many of which are pigments, and of polyphenols that are especially plentiful in brown algae.

Vitamins

Vegetables and fruits are an excellent source of many vitamins. They are the main suppliers of vitamin C (ascorbic acid), an important source of vitamin A and vitamin B_9 (folic acid), and a good source of vitamin E. Vitamin C is found especially in fruits, berries, and vegetables such as spinach and members of the cabbage family.

While vitamin A is most abundant in animal products, some plants, notably carrots, contain carotenoids (beta-carotene), which the body can convert to vitamin A. Vitamin E is derived mostly from plant oils, but also from members of the cabbage family, nuts, and oatmeal.

As mentioned above, vitamins A, C, and E are all antioxidants, with their content varying widely from one vegetable and fruit to another. The table on page 116 summarises some of the best plant sources.

Edible mushrooms are also a source of various vitamins, especially vitamin B, mostly in the form of B_2 and B_3. Those that have been exposed to some sunlight are also the only sources of vitamin D for vegans.

Seaweeds are a veritable treasure chest of vitamins, especially vitamins A, B (B_1, B_2, B_3, B_6, B_9), C, E, and K.

The vitamin B_{12} challenge

There are no eukaryotes, that is to say, humans and other animals, plants, mushrooms, or seaweeds, that can independently synthesise vitamin B_{12} (cobalamin) in a biologically active form. This is of no importance for plants and mushrooms as they have no need for vitamin B_{12}, and it is not found in a bioactive form in them. When cobalamin is found in a particular seaweed it is most probably in the inactive form, or else attributable to the seaweed's symbiotic relationship with the bacteria that produce it.

In the case of humans and other animals, the inability to synthesise vitamin B_{12} is problematic. It is a vital nutrient that plays a major role in the production of red blood cells and DNA, and the development of brain and nerve cells, as well as cognitive functions and speech in the case of humans. Even though there are bacteria in human intestines that can synthesise the active version of vitamin B_{12}, it is not absorbed because this process takes place in the large intestine, too late to be absorbed, and it is excreted.

Certain bacteria and other microorganisms biosynthesise vitamin B_{12} by a process of microbial fermentation. Recent research has also shown that cyanobacteria can synthesise it in a form, pseudo-cobalamin, that is sufficient for their own needs but biologically inactive in the human body. But it also seems that some microalgae can remodel this inactive form into an active one, thereby benefitting themselves and other organisms that derive B_{12} via their food chain.

Terrestrial animals are able to obtain vitamin B_{12} from bacteria that live in their digestive system or to a lesser extent those found in the soil, on grass, or in their fodder. Marine animals accumulate the vitamin from microorganisms such as plankton as they make their way up the food chain.

Humans, therefore, primarily get their vitamin B_{12} from animal sources—meat, seafood, dairy products, and eggs. Because of its role in the development of the brain, it is especially important that pregnant women have a sufficient intake of vitamin B_{12}, which can be stored in the liver. Consequently, one of the greatest challenges with following a purely vegetarian or vegan diet that excludes these foods is the risk of developing vitamin B_{12} deficiency. Unless such a diet includes certain species of seaweeds such as laver (*Porphyra/Pyropia* spp.) and fermented products, for example, yogurt or kimchi, made with particular strains of lactic acid bacteria, it has to be obtained from vitamin B_{12} supplements, which are often manufactured from bacteria. Another commonly used product is spirulina, made from dried cyanobacteria (*Spirulina platensis*), although it is not certain how much of its vitamin B_{12} content is in the biologically active form.

Vitamin	Examples of plants that are very good sources
A	Carrots, broccoli, peas, spinach, kale, Brussels sprouts, parsley, green beans
B (except B_{12})	Nuts, almonds, spinach, broccoli, red bell peppers, avocados, leeks, parsnips, kale, green beans, peas, garlic, lentils
C	Cauliflower, broccoli, bell peppers, parsley, Brussels sprouts, dill, spinach, radishes, leeks, tomatoes, potatoes
D	Mushrooms that have grown in sunlight
E	Olive oil, sunflower oil, corn oil, rapeseed oil, broccoli, Brussels sprouts, kale, nuts, almonds, oatmeal
K	Fava beans, olive oil, rapeseed oil, spinach, kale, broccoli, green beans, peas, parsley

Grandchildren are kind: a guide to a healthy diet

Mago wa yasashii is a Japanese mnemonic device first coined by Dr. Hiroyuki Yoshimura, a medical doctor and professor at Kanazawa University who died in 1998, as a reminder to eat a healthy diet to have a long life. The expression is made up by combining the initial syllables of traditional ingredients that represent seven important food groups: *ma* (*mame*, beans), *go* (*goma*, sesame seeds), *wa* (*wakame*, seaweed), *ya* (*yasai*, vegetable), *sa* (*sakana*, fish), *shi* (*shiitake*, mushroom), and *i* (*imo*, potato). It is easy to remember because it literally means 'grandchildren are kind.' Interestingly, the healthy, sustainable diet recommended by the EAT-Lancet Commission is made up of raw ingredients drawn from the same broad categories.

Micronutrients—minerals and trace elements

Vegetables and fruits have varying amounts of micronutrients, that is to say, minerals and trace elements.

Edible mushrooms are a good source of selenium, potassium, copper, iron, and phosphorus. The mineral content of seaweeds is typically ten times greater than that of terrestrial plants. Seaweeds have an abundance of iron, calcium, phosphorus, iodine, and magnesium. In particular, the brown algae from the Laminariales family, for example, *konbu* (kelp), sugar kelp, and tangle, can be very rich in iodine.

Seaweeds are also reasonable sources of trace elements, including zinc, copper, manganese, selenium, molybdenum, and chrome. As well, the potassium salt content in many species is greater than that of sodium salt. Using seaweeds to add the taste of salt to food confers a health advantage in that the overweight of potassium salt does not affect the sodium content in the blood. Consequently, it does not lead to an increase in blood pressure in the way that ordinary kitchen salt (NaCl) does.

Minerals and trace elements	Examples of plants where they are found in abundance
Calcium	Kale, broccoli, white and brown beans, oatmeal, parsley
Chrome	Kale, nuts, almonds, sesame seeds, soybeans
Copper	Seeds, nuts, almonds, whole grain products
Iodine	Lima beans
Iron	Spinach, almonds, lentils, pulses
Magnesium	Spinach, kale, bananas, avocados, chickpeas, broccoli, nuts, seeds, whole grain products
Manganese	Sweet potatoes, wheat germ, pine nuts, brown rice, lima beans, chick peas, spinach
Molybdenum	Lentils, yellow split peas, lima beans, kidney beans, soybeans
Phosphorus	Beans, seeds, nuts
Potassium	Beans, lentils, red beets, unpeeled potatoes
Selenium	Sunflower seeds, flax seeds, Brazil nuts
Sodium	Bok choi, artichokes, beets, carrots, celeriac, broccoli, cauliflower, Brussels sprouts
Zinc	Wheat germ, poppy seeds

What is the secret of Japanese longevity?

It is not unusual to read news articles about the longevity of Japanese people, whose life expectancy currently exceeds the global average by about twelve years. Within Japan, the inhabitants of the Okinawa Prefecture, a group of islands in the south, are notable for living to an extreme old age. There are five times as many centenarians, disproportionately women, as in the rest of the country. Their traditional diet consists mostly of locally sourced low-fat, low-salt foods, based on whole fruits and vegetables, legumes, tubers, tofu, and seaweeds. It is thought that the long life span on the islands is much more attributable to factors such as this low-calorie eating plan, along with an active lifestyle and a supportive community, than to genetic inheritance.

Small green Puy lentils.

The nutritional underpinnings of green cuisine

Natural poisons in plants and mushrooms

Wild plants and fungi may have poisons that are harmful to animals and people as a defence mechanism to avoid being eaten. These substances are often very bitter. The history of agriculture is, to a certain extent, an account of selective plant breeding to privilege varieties that pose no dangers and taste less bitter. In the effort to produce dependable crops that can grow quickly and be brought to market as cheaply as possible, the genes that code for characteristic aroma and taste substances have sometimes been lost. We earlier held up modern industrially grown tomatoes as a prime example of how aroma substances can fall by the wayside.

We often hear claims to the effect that it is healthier to eat produce from the heritage varieties that have been around for centuries and that these are in some undefined way more 'natural.' That this is not necessarily true can be seen from the following example of root vegetables such as parsnips, parsley root, and celeriac. All three have reasonable quantities of substances called psoralens, which are similar to coumarins and that, among other effects, elicit bitter tastes in cinnamon, tonka beans, strawberries, and cherries. Psoralens have been associated with increased risks of skin cancer if applied topically and coumarins can cause liver damage, but normally only if ingested in very large amounts. One can speculate that if these vegetables and fruits were now submitted to government health authorities as novel foods for human consumption, it is unlikely that they would be approved, even though they have been eaten quite safely for many centuries.

The best-known natural poisons in plants are alkaloids, which taste bitter and are harmful when ingested in large amounts. An example is solanine, which can be found in the fruits and tubers of members of the nightshade family, and that can act as a neurotoxin. That is why one should avoid eating potatoes that have turned green or unripe tomatoes. In the case of aubergines, it would be almost impossible to eat enough of them in one sitting to be at risk. Another bitter substance, caffeine, is a central nervous system stimulant and can have negative health impacts if consumed in large quantities. But we have become accustomed to it in moderation and enjoy its taste in coffee and tea.

Some cultivated plants have toxins that can be rendered harmless by cooking them before they are eaten. An important example is cassava (*Manihot esculenta*), a starchy tuber that is a major source of calories and carbohydrates for millions of people in tropical areas all around the world. As cassava contains dangerous levels of naturally occurring cyanide, however, the tubers must never be eaten raw and must be detoxified by cooking thoroughly in water or by fermentation. The pits of apricots, plums, peaches, and cherries and bitter almonds also contain a toxin that is broken down to cyanide by the digestive system. This is why it is dangerous to eat these fruits if their pits are broken or crushed as they can be lethal even in small quantities.

Cassava.

Many people are probably aware that certain beans, such as lima and kidney beans, cannot be eaten raw because they contain particular proteins, especially lectins, that can act as blood coagulants. Other toxins in beans can depress the natural activity of digestive enzymes and interfere with the absorption of important nutrients. Here again, cooking the beans over a long period of time neutralises these undesirable toxins.

Who has not experienced the special somewhat woolly, dry, and astringent sensation associated with eating rhubarb, spinach, and wood sorrel? It is caused by their oxalic acid content. The acid can be broken down by cooking the vegetables or by adding a little calcium carbonate or some dairy products to bind to it. Oxalic acid is toxic in the sense that it can suppress the absorption of calcium, which is vital for maintaining bone density. In addition, oxalic acid can increase the risk of developing kidney and gall stones.

Lectins, oxalic compounds, and various tannins are sometimes termed antinutrients because they may limit the absorption of important nutrients in the gastrointestinal tract.

Some mushrooms have poisons that are potentially lethal, and one should eat foraged ones only if they have been properly identified as safe and prepared properly. An example is fly agaric, an almost iconic species that is easily recognised from its white-spotted red cap and white gills. While they have psychoactive neurotoxins, these can be broken down by parboiling them twice in water. Commonly available cultivated mushrooms can also contain small quantities of various hydrazines, chemical compounds that are generally considered to be carcinogenic, but that can be removed by heating. It is, therefore, best to eat them fully cooked and in moderate quantities.

While wild plants and mushrooms generally harbour naturally occurring toxins, this is not the case with seaweeds. Very few of them are known to pose a danger for humans. Naturally, one must keep in mind that they should be harvested in clean water while still living in their natural habitat. On the other hand, microalgae are often toxic.

That being said, some species of seaweeds do contain substances that people are advised not to eat in too large quantities. For example, many brown algae from the order Laminariales, including sugar kelp, tangle, tangleweed, and *konbu*, are very rich in iodine. Even though a small, daily intake of iodine is absolutely necessary to ensure that the thyroid gland functions properly, an excess of this trace element in the food can lead to thyroid disorders. Other species, notably gulfweed (*hijiki*, *Hizikia fusiformis*), contain inorganic arsenic, which is undesirable even in small amounts and health authorities have warned against eating it. But the small quantities of a neurotoxic compound, kainic acid, found in some varities of dulse (*Palmaria palmata*) pose no danger to human health.

Microalgae are a different story. The two species, spirulina and *Chlorella*, described earlier are perfectly nutritious and edible. But there are approximately sixty to eighty marine microalgae that contain toxins that are potentially harmful. When these are ingested by animals such as shellfish, they bioaccumulate and can cause serious illness in humans who later eat them.

Raw, never cooked

A lifestyle choice that has gained some traction in recent years is raw foodism. There are two different philosophies within the movement, but their common mantra is that one must eat only food that is raw or has not been heated above a temperature of 40–48°C (104–118°F). In all cases, the ingredients are not 'prepared' in any traditional sense of cooking or frying, but may be fermented, dried, or aged. Some adherents follow a purely plant-based vegan diet consisting of fresh raw vegetables, fruits, nuts, seeds, sprouted grains, and legumes. Others also choose to add animal products such as dairy, eggs, meat, and fish, providing they can be eaten raw or processed without heat.

The movement seems to be based on promoting two concepts—that it is more authentic and natural for humans to eat raw food and, furthermore, that it is a healthy diet. That the first claim cannot be true is underscored by the reality that modern humans and their ancestors have been carnivores for at least two million years. That raw food can be a healthy option is a valid argument, but from an evolutionary perspective it is also a fact that *Homo sapiens* could not have developed and survived as a species on raw food alone. Our fertility during evolution would have been lessened if we had not had access to cooked food with a component of animal fat.

The temperature limit of 48°C (118°F) was chosen in order to preserve the enzymes in their active state in the raw ingredients. This is quite correct, but the possible health benefit of this type of preparation should be seen in the light of what happens when the food makes its journey from the mouth through the digestive system—enzymes and other proteins are broken down. It is also important to stress that the range of food that can safely be eaten raw might be lacking the full complement of nutrients, minerals, and vitamins needed for good health. It would be necessary to add vital macro- and micronutrients from other sources. Especially important are super-unsaturated fats (omega-3 fats primarily from sea food, such as fish and shellfish), vitamins (in particular D and B_{12}, only obtainable from animal products and algae), and minerals and trace elements (calcium, iron, selenium, and iodine).

Keeping those reservations in mind, one can benefit from embracing completely raw vegetables and fruits with their interesting textures—crispness, crunchiness, and juiciness—to complement our cooked food.

Plant-forward ingredients and food hygiene

Plants and mushrooms grow in dirt and seaweeds in water, both environments that are home to huge numbers of microorganisms. Many of these are harmless, but some can cause severe illness, for example, the *E. coli* bacteria that thrive in dirt and water contaminated with animal manure. Even a single bacterium can be the culprit and, given that we cannot see tiny bacteria and other microorganisms on the fresh ingredients, the solution is to pay a great deal of attention to keeping everything in the kitchen clean, especially knives, cutting boards, and storage containers. It is also important to be aware that bruised or damaged raw ingredients are more vulnerable to spoilage by undesirable bacteria than really fresh ones.

Refrigeration is a good safety measure to prevent bacterial growth. It goes without saying that it is important to wash the dirt off plants and mushrooms. And if seaweeds are to be eaten raw, they should first be washed in clean ocean water or lightly salted fresh water. Naturally, it is always a judgment call whether washing delicate ingredients such as salad leaves and berries does more harm than good but doing so carefully is advisable.

Preserving and fermenting

The vitamin and antioxidant contents of vegetables and their bioavailability depend largely on the specific vegetables in question and how they have been handled and prepared. Storing them for an extended period of time or subjecting them to heat will rob them of many of their vitamins. But as storage or heating may also increase the extent to which vitamins are absorbed during digestion, it is difficult to make a precise statement about their effect. Dehydrating, pickling, salting, and especially fermenting, will, however, greatly facilitate separating out the nutrients, including vitamins, from the plant matter as it is digested. Preserving vegetables without first cooking them in water is advantageous because it significantly limits the loss of water-soluble vitamins.

Preserved vegetables contain a variety of acids in different amounts, especially citric and malic acids, found naturally in the raw ingredients. In addition, acids such as vinegar and wine vinegar can be added to pickling liquids, and acetic and lactic acids can also result from fermentation. A certain level of acidity boosts the production of saliva and digestive juices, which helps to burn off carbohydrates and fats. Apart from that, the acidity is also an important aspect of what makes these conserved vegetables so palatable.

The combination of acid and salt helps to control the types of bacteria that can thrive on the vegetables. Acids can neutralise pathogenic bacteria and together with salt provide an environment that is favourable for the growth of new bacteria such as different types of lactic acid bacteria. Enzymatic activity is also affected by the acidity level, often slowing it down. As a result, nutrients are released over a longer period of time, which promotes better digestion.

Many claims are made about the health-enhancing properties of fermented and cultured foods and of the microorganisms that are used to prepare them. But few of these assertions are supported by scientific evidence. There are, however, three specific outcomes that are worth noting. Many people have said that they feel better, for example, with respect to their digestion, when they eat fermented foods. In addition, it is well known that fermentation leads to the formation and release of minerals, vitamins, and acids that are thought to promote better overall health. Other substances formed during fermentation can improve the taste of the food, having a beneficial effect by enhancing quality of life. Finally, some microorganisms are known to have inherent positive, probiotic properties.

There are many microorganisms and enzymes in our bodies that break down food into more readily bioavailable substances, for example, proteins, sugars, amino acids, and unbound minerals. Fermentation concentrates the nutritional value of a food. It can be seen as a process that decreases the burden on the digestive system and increases the uptake of nutrients—in a sense, 'pre-digesting' the food. Sometimes this involves enzymes and microorganisms that are able to do what our own digestive system cannot, that is, single out and release specific nutrients. Some of these processes can also break down toxins, for example, in cassava, so that the tuber becomes suitable for human consumption. In addition, fermentation can change the texture of certain foods, for instance, the tough parts of plants, leaving them softer and easier to chew.

Fermented foods often have an unpleasant smell. Whether or not one likes the aromas from a particular type of fermentation process is often culturally dependent. Fortunately, the fermentation processes also break down proteins into free amino acids and nucleic acids that are degraded to form free nucleotides, resulting in substances that impart umami and, frequently, synergistic umami. As a result, we find that the foods are palatable or at least have a taste to which we have become accustomed and that wins out over the off-putting smell.

The stomach 'talks' to the brain

Researchers have made some interesting discoveries about the way in which fermentation can have a positive impact on our overall health by helping to make raw ingredients tastier and more craveable but, paradoxically, also to regulate how much we eat. First of all, the process elicits umami, a taste that has been shown to stimulate the appetite and increase saliva flow. Saliva facilitates chewing and also contains substances that support the immune system by killing bacteria. Secondly, it has been found that there are dedicated glutamate receptors throughout the digestive system linked to a special communications axis between the brain and the stomach. Initially, the brain sends signals along this route to tell the stomach to release digestive enzymes. Later a message goes in the opposite direction to convey feelings of fulness and satiety from the stomach to the brain. In this way, the umami substances that were formed during fermentation work in seemingly contradictory ways that are also beneficial. They not only contribute to deliciousness and stimulate the appetite, but they also couple into a homeostatic mechanism that balances out the desire to eat with the need to limit food intake. These interactions are useful for those of us who have a tendency to overeat, while stimulation of the appetite helps those who have lost interest in food because they are sick or elderly and, consequently, become malnourished or weigh too little.

Tsukemono—a nutrient powerhouse

Tsukemono, the Japanese-style pickles described in the previous chapter, are an excellent example of how nutrients in a variety of vegetables can be preserved to enhance the content of important minerals, vitamins, soluble and insoluble dietary fibres, and macronutrients such as proteins and carbohydrates, that are present in the raw ingredients.

Tsukemono are prepared using about ten different techniques, described earlier, that may or may not involve fermentation. Regardless of whether or not the ingredients are fermented, however, they are not heated at any point. The conservation techniques have the effect of reducing the water content and lowering its activity. As a result, the concentration of nutrients, vitamins, and antioxidants will be relatively higher in the preserved vegetables than in their raw or cooked counterparts, but they can also be less bioavailable in the intestines. Fortunately, their natural enzymes, as well as those introduced from the fermentation media, if any, remain intact, allowing them to work on the vegetable solids to render the nutrients more accessible. It is important to keep in mind, however, that these enzymes eventually are broken down by those in the stomach and intestines as they pass through the digestive system.

These pickles are a very good source of certain vitamins. Lightly preserved ones can be rich in vitamins A and C. When they are stored over a longer period of time, however, the vitamin C content decreases gradually. *Kasu-zuke* have an abundance of vitamins B_1 and B_3 from the sake lees (*sake kazu*) that seeps into the vegetables and that can also be eaten if a little is left on the vegetables. *Nuka-zuke* absorb vitamins B_1 and B_3 from the fermented rice bran pickling medium. The vitamin B_1 content of *nuka-zuke* can be up to twelve times greater than that of the fresh vegetables from which they are made.

Reducing the water content of the vegetables also concentrates the dietary fibre so that it is three times greater than when they are in the raw state, which helps to ease the passage of food through the intestines. Finally, it is worth noting that the crunchy texture and crisp taste of *tsukemono* is an attribute that stimulates the appetite and releases saliva flow, thereby facilitating chewing, which is the first, vitally important step of the digestive process.

Preparing *tsukemono* can be summed up as a bit of a balancing act between using those techniques that will preserve the vital nutrients in the vegetables to the greatest extent possible and finding ways to make them more bioavailable on the one hand and ensuring that the still raw pickles are digestible on the other.

ORGANIC AND CONVENTIONAL FARMING

room for both

Over the past few decades, the supply of organic produce has increased markedly. At first this was driven by the growing awareness of the potentially harmful residues of substances such as pesticides and fungicides that might be found in plants and mushrooms grown according to commonly used agricultural practices.

More recently, issues related to climate change and sustainability have brought the consumption of healthy plant-based food into much sharper focus. But this is where things become complicated. First of all, the calculations required to quantify the difference between the relative footprint on the environment and impact on the climate of organic versus conventional agriculture are extremely complex. Secondly, there are questions as to whether we will be able to feed the growing world population using only our current organic practices. Thirdly, it may be difficult to reverse the prevalence of monoculture in favour of a diversified crop landscape. Finally, there is no consensus about whether or not to accept the cultivation of genetically modified crops designed to require less water, grow more quickly, and be more resistant to pests and fungi.

Is organic produce really better?

When it comes to sensory impressions, it is not a given that organic produce will taste either better than, or different from, its inorganic counterpart. Rather, the taste experience can easily be driven by how one feels about organic food in general or whether the vegetables and fruits look as attractive and uniform. Choice of cultivars can also have an effect as they may have a rustic appearance that has less eye appeal than their mass market cousins. Organic farmers often plant robust varieties that have naturally occurring substances that are toxic to pests and fungi, but this may affect taste, for example, increasing bitterness.

In terms of nutrition, there have been no definitive studies that indicate a clear health advantage in following a diet based on organic produce. Some research has shown that there is a small to moderate increase in some nutrients and omega-3 fatty acids. Other research has consistently reported that organic vegetables and fruits have a significantly higher content of antioxidant compounds such as anthocyanins, flavonoids, and carotenoids.

Food that is grown in countries where there is a strictly regulated system for certifying produce as 'organic' will probably show somewhat lower fertiliser, pesticide, and fungicide residues than conventionally raised ones. The differences may, however, be relatively small where there are governmental controls to ensure that residues are within limits that are considered safe. In many cases these limits have been lowered so that the produce that comes to market does not pose health risks. Farmers also have an economic incentive to reduce to the greatest extent possible the use of expensive synthetic fertilisers and pesticides in order to rein in their up-front costs.

Can enough food be grown using organic methods?

It is estimated that global food production will need to increase by as much as 60–100 per cent by 2050 to provide sufficient food for a population of 10 billion. This naturally raises the question of whether this is possible if there is a major move away from our current large-scale agribusiness methods. To a certain extent these methods have their roots in a time when agriculture evolved from being locally focused to taking advantage of better transportation options that enabled farmers to sell in distant markets. This could be done more efficiently and profitably by specialising and growing a single crop. A more recent development was the so-called Green Revolution that started in the 1960s and appeared to be the solution to the widespread famine that plagued many poorer areas of the world. It was based heavily on adopting technologies already in use in industrialised countries, especially reliance on chemical fertilisers and pesticides, selective plant breeding, and a degree of mechanisation. Land could be farmed more intensively because the application of synthetic fertilisers did away with allowing fields to lie fallow. While the yearly output of staple crops grew exponentially, this major shift away from traditional practices also had many negative consequences and is now being reassessed.

There is little argument about whether the present level of chemical and energy-intensive agriculture has had a major impact on the environment. It has been justly criticised for overdependence on monoculture that focuses on only one or two cultivars of particular crops selected for mass production. Limited crop rotation causes biodiversity loss and the concomitant loss of pollinators and beneficial insect predators. The run-off from synthetic pesticides and fertilisers can pollute waterways and the soil can become depleted. Organic farming is uniformly deemed to be a more sustainable approach that mitigates or avoids some of these problems.

Another aspect of the debate about the two approaches to food production centres on the greenhouse gas emissions linked to agriculture. An estimated 5 million hectares of forest, mostly in the tropics, are lost yearly to agricultural production, about half of them for cattle ranching and growing animal fodder. The use of heavy-duty agricultural equipment that burns fossil fuels adds to the problem. For some time, scientists have been stressing the incredible potential of planting new trees to remove a significant portion of the CO_2 that has accumulated in the atmosphere since the start of the Industrial Revolution. One way to take steps to make up for the lost trees is to adopt the practice of agroforestry in combination with organic land management practices. The basic idea is to plant trees and shrubs, often of many different species, in and around the fields where crops are grown. The aim is to harvest edible fruits from the trees while at the same time taking advantage of their ability to preserve topsoil and conserve ground water, all without the use of artificial fertilisers and extensive irrigation. This latter point is important given the droughts that have occurred with increasing frequency in agricultural areas. Furthermore, reducing the need to till the soil also helps to reduce soil erosion and use of machinery, while increasing retention of organic matter and nutrient cycling. By planting a variety of crops under the trees it is possible to create a type of vertically and ecologically integrated system that increases biodiversity and

that should result in greater yields with less labour and less recourse to machines to work the land.

One drawback, though, is that the yield from organic farming in developed countries tends to be considerably smaller than that from conventional agriculture, which would lead to pressure to expand the amount of land set aside for growing food. In highly industrialised areas this might not be possible. In less urbanised countries, however, organic farming methods can actually boost yields by growing crops on a rotating basis that can act as natural fertilisers and choosing cultivars that are less reliant on irrigation.

A convergence of organic and conventional farming may be underway

One of the issues that is sometimes associated with adopting a considerably more plant-based diet is whether or not it is necessary to rely exclusively on organically grown vegetables and fruits. Here cost can become a barrier. As discussed above, the readily available ones are more or less equivalent nutritionally, but they do not have the same cachet as organics. As a result, some people may feel that they do not fully support the health benefits that they hope to achieve by switching to a substantially plant-based diet.

Fortunately, there are some positive indications that we are in a period of transition and that organic and conventional farming practices are becoming less differentiated. This may partly be driven by the economics of farming. As organic agriculture becomes more and more mainstream, some large organic farms have, perhaps unwisely, adopted the monoculture model, presumably to achieve economies of scale. But cost factors are also motivating conventional farmers to look more closely at alternative methods.

Organic produce sells at a premium over most of the regular varieties and surveys indicate that this usually more than compensates for lower yields. Savings are to be had from reducing the use of synthetic fertilisers or replacing them entirely with manure and by planting cover crops that fix nitrogen and are then ploughed under to improve soil quality and prevent soil erosion. As a side benefit, this more diverse plant ecosystem attracts pollinators and beneficial insects. With vegetarian and vegan diets becoming more popular, especially from their being promoted by celebrities on social media, there is increasing demand. The trend seems to be moving predominantly in the direction of more sustainable agricultural practices.

FINAL WORDS

a way forward

In the Preface, we raised the question of whether there is a place for yet another book that advocates for a plant-forward approach to food preparation—what we call 'green cuisine.' As discussed below, the need to embrace eating habits that are more planet-friendly and promote better health is arguably one of the most consequential issues of the present time.

In order to address this challenge, we included significant background information about the physical nature of plants, especially their general lack of umami and sometimes unpalatable textures. These attributes often discourage some people, especially those whose kitchen cultures have traditionally relied on savoury animal-based dishes as major sources of protein, from exploring the full potential of plants, as well as seaweeds and fungi.

Although there is no need to be up to speed on the underlying science in order to prepare the recipes in the book, it can be helpful when adapting them to suit individual preferences, substituting ingredients, or taking them in other directions. It works in tandem with an effort to expand our culinary horizons by presenting some interesting and possibly novel options for putting together meals that are healthy, delicious, utterly craveable, and at the same time more sustainable. Turning to plant-forward foods is a way to address two existential problems, namely, the impact of what we eat on the environment and on health outcomes.

The average diet that is prevalent in many parts of the world, particularly in the better-off countries, which incorporates significant portions of meat and dairy products, is now routinely described as unsustainable, unhealthy, and a threat to the survival of humanity and the planet. In response to the still increasing global demand for meat, its production has tripled over the past fifty years. Unfortunately, animal husbandry places an outsized burden on the environment by using plants as fodder rather than as primary sources of nutrition. The resulting inefficient conversion of plant protein to animal protein is extremely wasteful. Diverting these crops for human consumption instead could help to alleviate severe hunger in areas of food insufficiency and thereby reduce the incidence of serious diseases and malnutrition.

Raising livestock also accounts for the lion's share of deforestation, and the greenhouse gas emissions released as part of the animals' digestive processes are estimated to make up about one sixth of the total attributable to anthropogenic activities. Letting forests grow and act as carbon sinks instead of turning them into pastureland would have a net positive effect on climate change. The market for animal protein has also led to commercial overfishing of some marine stocks almost to the point of depletion, with a concomitant loss of biodiversity and degradation of the ocean environment.

One response to the demand for animal protein is to replace it with products that mimic the taste and texture of meat and cheese, but this is a step backwards. Many of these are made by extracting or isolating proteins from peas and soybeans using highly developed industrial techniques, which incur significant energy costs and waste the remaining edible parts of the legumes. In addition, the long-term health effects of their additive contents and the way in which these ultra-processed foods are manufactured are as yet unknown. These products are generally more expensive than the ones that they are meant to replace. A better alternative is for people to become more informed consumers and to seek out and support farming and food processing practices that have a smaller ecological footprint.

Doing so also has the potential to reduce the waste that is another culprit contributing to making our present food system unsustainable.

A second aspect of some current eating patterns is that there is an overreliance on ready-to-eat foods and snacks. Apart from their appeal as convenient time-savers, they are, in many cases, ultra-processed in order to achieve tastes and textures that make them almost irresistible. Their lists of ingredients include many additives with complex chemical compositions that are not items that would be found in an average household and many have high levels of fat, salt, and sugar. The consensus conclusion of numerous studies of these foods is that they pose serious risks of adverse health outcomes, including increases in the incidence of cardiovascular disease, mental health disorders, obesity, and type 2 diabetes. While it is not possible because of ethical issues or for practical reasons to conduct longitudinal, controlled studies of these effects, the epidemiological evidence all points, to varying degrees, in the same direction.

Both of these types of food choices present identifiable obstacles to finding a way forward. Thanks to their evolutionary path, humans became meat-eating omnivores and also developed a preference for food that has umami, sweetness, and a pleasing texture. This inevitably leads to the question of the best and most feasible way to encourage people to alter their deeply ingrained eating patterns in favour of ones that place fewer demands on increasingly scarce resources and are healthier.

Mitigating the impact of current dietary patterns in order to ensure that there is a fair distribution of food for everyone is a matter of great urgency now and will become even more so in the immediate future. This may require people of all ages in higher income countries to make significant adjustments to what they eat and should ideally be accompanied by reaching out to involve children and youth, those who will be most affected by the failure to do so. Given that what they learn to eat at home in their early years can become deeply embedded in their food preferences, it is important to start them off on a healthy, sustainable pathway. These actions taken at the level of the individual household can be reinforced and amplified by advocating for educational programmes that teach food science and promote future-oriented eating habits, as well as for making plant-based foods available in public settings.

History teaches us that unless there are external events that act as a force majeure, it is difficult, if not almost impossible, to make these changes over a short period of time. Nevertheless, a viable and realistic option is to transition to a modified diet in stages that stop short of being fully vegetarian or vegan. An initial step is to adopt what is known as a 'whole foods' approach. It is based on starting with ingredients in their original state and adding only simple seasonings that have no artificial components. This can be coupled with embracing flexitarian ideas that call for the inclusion of small amounts of umami-rich ingredients of animal origin—seafood, meat, eggs, or dairy products—almost as condiments, to enrich the flavour of dishes made primarily with plants, seaweeds, and fungi. These can easily be paired with the Flavour Accelerators to create synergy and enhance the overall effect. Putting the focus on the sensory perception of a dish may encourage many more people to increase their intake of green cuisine without compromising the taste experience and their enjoyment of the food.

It is necessary to acknowledge that effecting these changes may not be an easy task for many who already feel time-stressed and may find it difficult to make the effort to cook from scratch. If they lack the necessary culinary skills they may need to acquire them, as well as knowledge about how to combine ingredients to prepare plant-forward dishes that are as appealing as those that have animal products as their focal point. Nevertheless, this is an essential aspect of green cuisine. The reward lies in knowing exactly what is in the food and having control over the fat, salt, and sugar content.

Applying flexitarian concepts in the transition to a plant-forward culinary culture may need to be eased in over a longer time frame, depending on individual circumstances and public policy choices, in order to have a lasting impact. Remaining faithful to it requires a constant, consistent effort and a willingness to adapt to this new reality. The important point is to embark on the journey and to stay the course—one dish at a time.

A GLOSSARY OF THE MAIN PLANT INGREDIENTS FROM A TO Z

characteristic tastes, aromas, and textures

It is very rare that a given raw ingredient is characterised solely by its taste. For example, it can be difficult to identify a herb correctly if one plugs one's nose and covers the eyes so as to be unable to smell its aroma substances or see its leaves. But taste and mouthfeel together with some characteristic aroma substances can help to identify the raw ingredient that one is sampling. Conversely, it is easy to tell whether one has taken a sip of wine or of wine vinegar, even though both are acidic.

The aroma profile of all types of food, including vegetables and fruits, is characterised by an enormous number and variety of aroma substances. Many of the substances are generic and are common to many different foods. In some cases, the characteristic smell of an ingredient is determined by only a small number of these substances, but things are usually much more complicated. What this means is that it is difficult to recreate artificially the aroma of a particular raw ingredient. Even though enough of the major components have been captured to enable one to figure out to which ingredient it relates, one often concludes that something is missing without being able to clearly identify it.

In the case of vegetables, the combination of aroma and taste substances is not always sufficient to allow one to tell them apart. It is easy to demonstrate this by manipulating the colour and mouthfeel. A classic experiment involves members of the cabbage family, for example, cone cabbage and broccoli. If each of them is reduced separately to a purée, dyed red, and set with a suitable gelation agent, it has been shown that only 5 per cent of a group of tasters can determine what the original raw ingredient was based only on the taste and smell of the gels.

The raw ingredients described in this section are primarily ones that we think of in connection with savoury dishes. We have chosen not to incorporate fruits, herbs, grains, and seeds in a systematic fashion, although a few are included. The common names of the ingredients can vary from the ones used in other contexts, publications, and geographical locations. This is especially true of different varieties of the same species.

In order to be precise, the raw ingredients are identified by their scientific names, following the standard binomial nomenclature of genus followed by species. Information about the many possible varieties and hybrids of each species is given only in

some cases. Abbreviations used are:
- sp. – species when not specified
- spp. – several species from the same genus
- ssp. – subspecies
- var. – variety (type, cultivar)
- convar. – group of varieties

For detailed information about the nutritional content and dietary contribution of the raw ingredients, readers should seek out appropriate reference literature.

Apples

Apples (*Malus* spp.) belong to the rose family (Rosaceae) and are the third most important fruit crop worldwide. It is estimated that there are about 7,500 cultivars of eating apples (*Malus domestica*). New hybrids are being introduced regularly, allowing consumers to home in on the characteristics that they consider most important.

Taste: Ranges from very tart, almost sour, to mild and sweet.
Aroma: Derived from volatile organic compounds and depends on the variety.
Texture: Crisp and juicy when freshly picked; can be mealy when mature and old.
Culinary applications: Apples are generally associated with sweet dishes and baked goods, but can also be treated like vegetables, for example, as a raw ingredient in salads. Pickled ones are served as versatile, sweet and sour condiments for savoury dishes.

Artichokes

Artichokes (*Cynara cardunculus* var. *scolymus*) belong to the aster family (Asteraceae) and are related to thistles. The edible portion, which is globe-shaped, sits at the end of a tall, thick stalk. It is covered with many small flower buds in the shape of triangular green scales and their bracts surrounding a fleshy base, known as the heart. The fleshy lower parts of the bracts and the heart are eaten before the plant starts to bloom. Artichokes are sold fresh and artichoke hearts also preserved in water or oil in tins and jars.

Taste: Somewhat sweet, a little bitter, slightly nutty.
Aroma: Earthy.
Texture: Bracts are a bit crunchy and crisp; the hearts are buttery and soft.
Culinary applications: Artichokes can be eaten raw in salads or roasted, boiled, included in stews, made into soups, sauces, and dips, and used as pizza toppings.

Arugula

Arugula (*Eruca vesicaria* ssp. *sativa*), also called rocket, belongs to the cabbage and mustard family (Brassicaceae). It has small, bitter pinnate leaves and is a popular salad vegetable.

Taste: Peppery, bitter, mustard-like.
Aroma: Spicy, peppery, nutty.
Texture: Crisp and juicy.
Culinary applications: Arugula leaves are eaten raw in salads. They can also be wilted or sautéed like spinach for use in warm dishes and made into pesto.

Asparagus

Common asparagus (*Asparagus officinalis*) belong to the asparagus family (Asparagaceae). The edible parts of the plant are in the shape of long, tender shoots, usually called spears, with a small scale-like leaves at the top and come in green, white, and purple varieties. The colour of purple asparagus is due to anthocyanins. Their rootstocks are identical and while they are growing underground, they have no chloroplasts and are white. Once the shoots emerge from the soil, they turn green or purple because the plants are now producing chloroplasts in order to photosynthesise. Others are prevented from changing colour by mounding earth to cover the spears as they continue to grow.

Taste: Moderately salty, sour, and umami tastes when fresh. Bitterness comes from saponins, which are also found in peas and beans, and sulfurous tastes from dimethyl sulfide, a substance also given off by boiled milk and fish that is no longer fresh. The green varieties have more umami and a stronger, slightly more grassy taste than the mild white ones. The purple cultivars are a little nutty and sweet.
Aroma: Mildly sulfurous aroma due to asparagusic acid.
Texture: Mouthfeel is the most important aspect of the sensory experience of eating asparagus. Their texture is generally crisp, and fresh spears can easily be snapped in two. Young white asparagus are especially succulent and buttery. But the mouthfeel can be completely ruined when the spears become woody if they are left for too long after being harvested or if they are overcooked and turn mushy.
Culinary applications: Asparagus can be eaten raw, steamed, boiled, roasted, pickled, and fermented. Fresh peeled white asparagus can also be immersed in an umami-rich marinade with, for example, *miso* or soy sauce.

Aubergines

Aubergines, also known as eggplants, are the fruits of the *Solanum melongena* plant, which is a member of the nightshade family (Solanaceae) and related to potatoes and tomatoes. While low in calories, they are fibre-rich and a source of important vitamins, minerals, and antioxidants. They are found in many varieties, sizes, and shapes, for example, round, pear-shaped, or elongated. Most have a deep purple or bluish skin, but some are striped, white, or green, such as the highly prized *ao-daimaru*, one of the traditional Kyoto vegetables. Japanese cultivars have a thin skin, which makes them ideal for pickling. Aubergines are eaten whole, either with skin or peeled; the flesh is spongy and has small seeds. The empty space between the cells collapses when aubergines are cooked or grilled, and they lose volume but readily absorb fats and aromatics. If they are rubbed with salt to draw the liquid in the cells out into the spaces between them before deep-frying, they will absorb less fat.

Taste: Slightly bitter and astringent tastes can be lessened by rubbing the aubergines with salt before cooking, which brings out sweetness and more complex tastes.
Aroma: Earthy aroma when fresh; will give off a sulfurous smell if spoiled.
Texture: Texture can be creamy or meaty depending on how they are prepared, while old ones may be woody.
Culinary applications: Aubergines are featured in many food cultures in savoury side and main dishes and sauces. They can be grilled, simmered, roasted, fried, and puréed, as well as pickled to make *tsukemono*. They are an important ingredient in ratatouille and grilled slices are sometimes used as a meat substitute by vegans.

Avocados

Avocados are fruits of a small, evergreen tree (*Persea americana*), which belongs to the laurel family (Lauraceae). The fruits are pear-shaped, have a tough, often hard, skin, and a large pit. The flesh is characterised by its up to 30 per cent fat content. About 84 per cent of the fats are unsaturated, mostly monounsaturated oils, but there are also smaller quantities of omega-3 fatty acids. They contain little sugar or starch and no sodium or cholesterol, but are good sources of fibre and important nutrients, including vitamin K and potassium. Although there are many different cultivars, the Hass variety accounts for about 80 per cent of the entire crop. All of these trees are descended from a single 'mother tree' that was raised and patented by Rudolph Hass in California in 1935.

Taste: Somewhat bland, soapy, a little nut-like.
Aroma: Aromas are due to slightly spicy terpenes, for example, caryophyllene, as well as breakdown products of fatty acids with seven to ten carbon atoms.
Texture: Highly rated soft, buttery, and creamy texture.
Culinary applications: Raw avocados are used in salads and dressings, mashed to make guacamole, or served on toast. They are also incorporated into smoothies and can be used in sweet baked goods and desserts.

Beans

There are very many different varieties of beans, which are legumes and belong to the pea family (Fabaceae). Although botanically speaking beans are fruits, they are usually thought of as vegetables. Beans are harvested when their enclosing pods are fully grown. They can be broadly divided into two different types—those that are eaten whole or shelled while fresh and those that are usually shelled, then dried, and rehydrated before cooking. The best-known varieties that are eaten fresh are common beans (*Phaseolus vulgaris*), which can be green, yellow, or purple, flat Romano beans (*Phaseolus vulgaris* var. 'Hilda'), and runner beans (*Phaseolus coccineus*). Purple beans turn green when boiled with a little acid. Fava beans (*Vicia faba*) and soybeans (*Glycine max*) are sometimes sold fresh in their pods, but these must be removed before cooking. Many bean varieties, for example, borlotti, lima, mung, flageolet, lupini, azuki, and kidney, are available in dried form year-round and are soaked and boiled before use. They can also be bought ready for use in cans. The seed coat of some types can be tough and bitter and is removed from the individual seeds. Some beans contain a toxic compound, lectin, that must be broken down by cooking.

Taste: Fresh green beans taste slightly bitter and yellow wax beans are similar but milder. The taste of other types of beans varies greatly, for example, kidney beans have a robust, heavy taste, while flageolets are mild and delicate, and azuki beans are sweet enough to be used in desserts.
Aroma: Fresh green beans have a herbal aroma, while the aromas of dried beans are linked to specific varieties.
Texture: Depends to a great extent on how they are prepared. Fresh beans are crisp and crunchy but turn mushy if overcooked. Texture of previously soaked dried beans depends on cooking time.

Culinary applications: Fresh beans are generally steamed or roasted and served as a vegetable dish or added to salads. Dehydrated ones are boiled before use in other dishes, such as salads, soups, and stews. All types of beans can be sprouted, fermented, or pickled.

Beets

Beets (*Beta vulgaris* convar. *crassa* var. *conditiva*) belong to the amaranth family (Amaranthaceae) and are related to spinach, chard, and quinoa, among others. The taproots, stems, and the leaves are all edible. Most beets are reddish-purple in colour, although there are also golden yellow varieties. The characteristic pigments responsible for these colours are betalains and betaxanthins, respectively. Betalains are affected by heat and are water soluble. This is why some of the colour seeps out into the cooking water.

Taste: Earthy, slightly bitter taste, which becomes sweeter when they are roasted.
Aroma: Earthy aroma due to geosmin, a volatile terpene, which partly disappears during the cooking process.
Texture: Juicy and crisp when raw; soft and succulent when cooked; may be woody when old.
Culinary applications: Beetroots can be eaten raw in salads, boiled, pickled, and toasted to make chips. They are often spiralised for use as decoration or to make vegetable pasta. The taste, aroma, and texture of the leaves are similar to those of spinach and chard, and they can be used interchangeably.

Belgian endives

Belgian endives (*Cichorium intybus* var. *foliosum*), also called chicory or witloof, are a type of chicory belonging to the aster family (Asteraceae). The leaves form a small head with an elongated pointed shape. Their shoots are white as long they are in cold soil and completely in the dark, causing them to be very bitter. As soon as the plants are exposed to light, they start to turn green at the tips and lose much of the bitterness

Taste: Bitter, strong, toasty, nutty.
Aroma: Mild.
Texture: Crisp and juicy.
Culinary applications: Belgian endives are often eaten raw in salads and can be pickled. When grilled, they become sweeter.

Bell peppers

Bell peppers (*Capsicum annuum*), also called capsicum, are fruits and, like chillis, are members of the *Capsicum* genus and belong to the nightshade family (Solanaceae). They come in many shapes and sizes, including bell-shaped, round, and elongated, and are hollow with seeds inside. All varieties are green at first and some change colour as the fruits mature. The most common varieties are green, yellow, orange, or red, when fully ripe. Less common varieties can be off-white, purple, or brown.

Taste: Generally sweet. In contrast to other members of the genus, never have a sharp, burning taste because they lack capsaicin, an irritating substance.
Aroma: Sweet herbal smell due to unsaturated aldehydes.
Texture: Crisp and crunchy when raw; soft when cooked.
Culinary applications: Bell peppers are often eaten raw in salads. They can be grilled, baked, and roasted, as well as stuffed or added to sauces and savoury breads.

Black salsify

Black salsify (*Scozonera hispanica*) belongs to the aster family (Asteraceae). It has a long edible taproot that is brownish-black on the outside and chalk white inside. When peeled, the flesh discolours very quickly if it is not immersed in a mild acidic solution. The leaves are also edible.

Taste: Sometimes called 'oyster plant' because its taste can be compared to that of oysters and seafood, with hints of liquorice and umami.
Aroma: Fairly mild smell, somewhat like that of artichokes.
Texture: Dense, firm, crisp, nutty, and crunchy.
Culinary applications: Salsify must be peeled and is eaten raw in salads, boiled, steamed, glazed, or grilled.

Bok choy

Bok choy (*Brassica rapa* var. *chinensis*) belongs to the cabbage and mustard family (Brassicaceae) and is related to yellow mustard, Chinese cabbage, and turnips. The plant does not form heads but has a bulbous bottom with green leaves.

Taste: Slightly peppery and somewhat sweet taste, less strong than that of spinach but more intense than that of water chestnuts.
Aroma: Slightly sulfurous aroma typical of cabbage.
Texture: Crisp and juicy when raw; soft when cooked.
Culinary applications: Bok choy is eaten raw in salads, steamed, or stir-fried. It can be simmered in *dashi* or *ponzu* and is ideal for making *koji-zuke*.

Broccoli

Broccoli (*Brassica oleraceae* convar. *botrytis* var. *italica*) is a green plant that belongs to the cabbage and mustard family (Brassicacea). The immature florets of its large, flowering head, the stalk, and small leaves are all edible.

Taste: Slightly sweet and slightly bitter when raw; milder when cooked.
Aroma: Typical cabbage aromas are due to sulfurous compounds.
Texture: Normally crisp; the stalks can become woody if they have been left too long after harvest.
Culinary applications: The florets can be served raw as crudités. Boiling, steaming, roasting, or deep-frying removes some of the sharp, irritating cabbage taste. Broccoli is suitable for *koji-zuke* and the stalks can be fermented.

Broccolini

Broccolini is the trade-marked name for a hybrid developed in Japan in 1993. It is a cross between broccoli (*Brassica oleracea* convar. *botrytis* var. *italica*) and Chinese broccoli (*gai lan*) (*Brassica oleracea* var. *alboglabra*), both of which are members of the extensive cabbage and mustard family (Brassicaceae). The flower heads are smaller than those on broccoli, while the stems are relatively longer and thinner and are usually eaten as well.

Taste: Bitter, slightly irritating taste, but milder than that of broccoli.
Aroma: Sulfurous aromas typical of the cabbage genus.
Texture: Crisp if not overcooked.
Culinary applications: Broccolini can be eaten raw, boiled, roasted, or baked. Preparing them like *tsukemono* with *koji* elicits sweetness and umami and lessens the bitterness.

Brussels sprouts

Brussels sprouts (*Brassica oleracea* convar. *fruticosa* var. *gemmifera*) are the small edible buds that grow along the tall stalk on a plant belonging to the cabbage and mustard family (Brassicaceae).

Taste: Somewhat bitter when raw; become sweeter when cooked.
Aroma: Sulfurous aromas.
Texture: Crisp when raw; become softer depending on how they are cooked.
Culinary applications: Individual leaves are sometimes separated and eaten raw in salads. The sprouts can also be boiled, sautéed, stewed, steamed, or braised, which results in a deeper, heavy cabbage taste and aroma; often used to add visual appeal to a dish.

Cabbages

What are generally called cabbages are a cultivar group belonging to the cabbage and mustard family (Brassicaceae). They are made up of a compact head of thick, alternating leaves that can be smooth, wavy, or crinkly and that may be green, white, or purple. The best-known varieties are: green cabbage (*Brassica oleracea* convar. *capitata* var. *alba*), also known as white cabbage; red cabbage (*Brassica oleracea* convar. *capitata* var. *rubra*), which has attractive deep purple leaves with veins that are white on the inside and add visual appeal to a dish; savoy cabbage (*Brassica oleracea* convar. *capitata* var. *sabauda*), which has pale green crinkly leaves; and cone cabbage (*Brassica oleracea* convar. *capitata* var. *conica*), which has a distinctive conical shape and comes in green, purple, and red varieties.

Taste: Varies from one type to another but generally somewhat bitter because of the polyphenol content and irritating on account of the enzymatic breakdown of glucosinolates to form isothiocyanates.
Aroma: Strong aroma, which becomes more noticeable after the leaves are cut and cooked, due to an organic sulfurous compound, 5-methylthiopentanenitrile.
Texture: Crisp and juicy unless old or overcooked.
Culinary applications: Cabbages are a staple ingredient in many traditional kitchen cultures. They can be eaten raw, especially in coleslaw or as a salad, used in soups, boiled, marinated, fermented, braised, baked, grilled, or stewed.

Carrots

Carrots (*Daucus carota* ssp. *sativus*) belong to the umbellifer family (Apiaceae) along with celery, parsley, and dill. The taproots of the plant are among the most commonly eaten vegetables. Although they are often associated with the colour orange, there are red, yellow, purple, and white varieties, with the last being the most aromatic.

Taste: Slightly bitter when raw; can become sweet and even very sweet when cooked.
Aroma: Notes of pine due to terpenes from the aromatic substances that are concentrated in the outer layer of the root.
Texture: Crisp and crunchy when raw; soft, and slightly creamy when cooked; can become woody with age.
Culinary applications: Carrots can be eaten raw and prepared in virtually all other ways. They are also a popular ingredient in baked goods

A glossary of the main plant ingredients from A to Z

Cauliflowers

Cauliflowers (*Brassica oleraceae* convar. *botrytis* var. *botrytis*) belong to the cabbage and mustard family (Brassicaceae) and are in the shape of a compact head made up of florets. Generally, only the florets are used, although the stalks can be sliced thinly lengthwise and eaten raw and the leaves can be cooked. Cauliflowers come in white, green, orange, and purple varieties. A special variety, Romanesco, is chartreuse-coloured and has characteristic pointed florets. Because the florets are not mature, they are less fibrous than the stalks and contain a great deal of pectin and hemicellulose. As a result, the florets can be puréed to a creamy consistency. For this reason, it is easy to overcook cauliflower and it becomes mushy.

Taste: Mildest taste of all cabbage types, ranging from sweet to slightly bitter with notes of pepper and nuts.
Aroma: Somewhat sulfurous, but milder than that of green cabbage.
Texture: Stalks are meaty; florets are firm and crisp when raw; very creamy when cooked or puréed.
Culinary applications: The florets are often served raw as crudités and in salads. They can be boiled, steamed, grilled, roasted, fried, and pickled.

Celeriac

Celeriac (*Apium graveolens* var. *rapaceum*) is closely related to celery (*Apium graveolens* var. *dulce*) and is a member of the umbellifer family (Apiaceae). It grows in the form of a large knob that is not a root but a swollen part of the stem and is often a little woody. Celeriac contains only about 5 per cent starch.

Taste: Earthy, nutty, and slightly sweet, but can be sharp, bitter, and acidy.
Aroma: Herbal aroma from phthalides.
Texture: Firm, crisp, and woody when raw; creamy when cooked.
Culinary applications: Celeriac must be peeled and then can be cut up or grated and used as a root vegetable, either raw or cooked. It oxidises quickly and the pieces should be immersed in water with a little lemon juice or vinegar to prevent browning if they are to be used raw. It is also sliced and simmered in *dashi*, or breaded and fried to make celeriac burgers. The small, delicate leaves, like those of celery tops, can be used to add taste in a bouquet garni.

Celery

Celery (*Apium graveolens* var. *dulce*) is closely related to celeriac (*Apium graveolens* var. *rapaceum*) and is in the umbellifer family (Apiaceae) as are carrots and fennel. Both stalks and leaves can be eaten.

Taste: Leaves are slightly bland; stalks are a little bitter and salty.
Aroma: Herbal aroma from phthalides.
Texture: Crisp and juicy; the stalks are a bit woody and stringy due to the xylem.
Culinary applications: The stalks are eaten raw as crudités and in salads, for example, Waldorf salad, and added to soups. The celery leaves can be used as part of a bouquet garni.

Chickpeas

Chickpeas are the seeds of an annual legume (*Cicer arietinum*) that belongs to the pea family (Fabaceae). They are available in dried form or ready to use in cans, and ground into flour. Chickpeas are very protein-rich, and they easily become mushy if overcooked. The cooking water (*aquafaba*) contains some proteins and carbohydrates and, like egg whites, can be beaten to a foam and used as an emulsifier that can be substituted for eggs in mayonnaise and in vegan dishes. Dried chickpeas must always be rehydrated and cooked before use.

Taste: Somewhat sweet taste, also described as mild, nutty, and earthy.
Aroma: Mild when cooked; gives off an offensive odour while being soaked.
Texture: Depending on cooking time, a little firm or mealy.
Culinary applications: Whole cooked chickpeas are added to salads, stews, and curries, and puréed to make falafel, hummus, and pâtés.

Chinese cabbage

Chinese cabbage (*Brassica rapa* ssp. *pekinensis*), also called napa cabbage, belongs to the cabbage and mustard family (Brassicaceae). It grows in the shape of an elongated head made up of loosely packed, pale green crinkly leaves.

Taste: Considerably milder and sweeter than green cabbage.
Aroma: Sulfurous, less strong than that of green cabbage.
Texture: Delicate but still crisp.
Culinary applications: Chinese cabbage is eaten raw in salads, stir-fried, and fermented, for example, to make *kimchi*.

Chives

Chives (*Allium schoenoprasum*) are a variety of onions, a genus that belongs to the amaryllis family (Amaryllidaceae). The herb is made up of small, edible tube-shaped stems, some of which are topped with an edible flower.

Taste: Very mild onion taste.

Aroma: Mild volatile sulfurous aromas released when cut.
Texture: Slightly crisp.
Culinary applications: Chive stems are used primarily as herbs to add a mild onion flavour or a touch of green colour. The flowers are tossed into a salad, added to herb infused oils and vinegars, and used as garnishes.

Collard greens

Collard greens (*Brassica oleracea* var. *viridis*) belong to the cabbage and mustard family (Brassicaceae). The plants have large, thick, dark-green leaves that do not form a head.

Taste: Bitter when raw; mellower and a little earthy when cooked.
Aroma: Typical sulfurous cabbage family aromas.
Texture: Chewy and tough when raw; soft and mushy after slow cooking.
Culinary applications: Collard greens are usually boiled, sautéed, stewed, or fermented.

Corn (maize)

Corn (*Zea mays*) is a member of the grass family (Poaceae) and is the most widely grown staple food crop in the world. There are six major varieties of this cereal grain, which comes in several colours and different degrees of sweetness. The individual kernels are actually fruits but are most often eaten as a vegetable.

Taste: Sweet and umami.
Aroma: Sulfurous aroma, for example, from dimethyl sulfide, when cooked and notes of caramel when roasted.
Texture: Soft and slightly creamy when cooked. Small, immature corn ears are crisp.
Culinary applications: Corn can be eaten raw, but both the baby and mature cobs are generally boiled or grilled. The kernels are also ground into meal and flour, which is used in baked goods, or made into chips, cereals, and tortillas.

Cucumbers

Common cucumbers (*Cucumis sativus*) belong to the gourd family (Cucurbitaceae) and are annual summer vegetables found in many varieties, including slicing or garden cucumbers, English and Persian cucumbers, and small ones for pickling. Cucumbers have a significant water component, about 95 per cent, and contain only a few essential nutrients, but have reasonable quantities of vitamin K. Cornichons, also called gherkins, are pickled cucumbers from the *Cucumis anguria* species, which are quite small and have a nubbly skin.

Taste: Generally mild, slightly sweet taste; can be bitter due to the presence of cucurbitacin, a triterpene, if grown under adverse conditions.
Aroma: Dominated by unsaturated aldehydes, especially (E,Z)-nona-2,6-nondienal and (E)-2-nonenal, that are formed enzymatically from unsaturated fatty acids (linoleic and linolenic acids) when the cells of the cucumber are damaged, for example, by crushing it or cutting into it. Aroma substances can be extracted in oil.
Texture: Thin slices are generally crisp, while thicker pieces are crunchy. Dehydrating amplifies these qualities.
Culinary applications: Cucumbers are often eaten raw in salads or used to make cold soups; pickled, for example, in a salt, sugar, and vinegar solution (cucumber salad); or dehydrated to make *tsukemono*.

Daikon

Daikon (*Raphanus sativus* var. *longipinnatus* or var. *acanthiformis*), also known as Chinese radish, belongs to the cabbage and mustard family (Brassicaceae) as do other types of radishes. Unfortunately, this large, elongated white taproot is overlooked in many food cultures.

Taste: Fairly strong, sharp, peppery, and biting taste that is less intense after cooking.
Aroma: Mild when whole but has a distinct pungent aroma due to the presence of sulfur-containing compounds when cut or grated.
Texture: Crisp when raw or pickled; becomes soft and succulent when simmered or boiled.
Culinary applications: *Daikon* can be eaten raw in salads, simmered, boiled, pickled, dehydrated, marinated, and fermented to make *tsukemono*.

Fennel

Fennel (*Foeniculum vulgare* var. *azoricum*) belongs to the umbellifer family (Apiaceae) and is related to carrots, parsley, dill, and celery. It has a swollen bulb-like stem base that together with the stalks and leaves is eaten as a vegetable. The seeds are aromatic and used as a spice.

Taste: Mild anise taste, similar to that of liquorice.
Aroma: Liquorice-like due mainly to anethole, an essential oil also found in anise and star anise. Other aromatics are limonen (citrus), estragol (tarragon), and fenchone (camphor).
Texture: Crisp, crunchy, juicy, and slightly woody when raw.
Culinary applications: Fennel bulbs are often eaten raw in salads, and can be steamed sautéed, stewed, or grilled. The leaves are used as fresh herbs.

Figs

Figs are small edible tear-shaped fruits from the fig tree (*Ficus carica*), which belongs to the Moraceae family. They have a thin skin covering a fleshy interior containing tiny, crunchy seeds and are eaten both fresh and dried. They are green when unripe and turn brown or purple when mature.

Taste: Slightly sour and bitter when unripe; sweet when ripe.
Aroma: Special spicy aroma of ripe figs due to linalool, a floral and spicy terpene.
Texture: Soft on the outside and grainy inside when ripe; fragile and easily damaged.
Culinary applications: Unripe figs can be eaten raw as vegetables in salads and as a garnish. They are also pickled and marinated. Ripe figs are eaten as dessert fruits, often with cheese, and flambéed. When dried they are eaten as a snack and can be used in baked goods.

Gai lan

Gai-lan (*Brassica oleracea* var. *alboglabra*), also called Chinese broccoli, belong to the cabbage and mustard family (Brassicaceae). The plants have long stems with glossy blue-green leaves and small florets.

Taste: Sweet stalks and slightly bitter florets.
Aroma: Sulfurous aromas typical of the cabbage family.
Texture: Leaves are soft and similar to spinach, while the stems are crisp and firm.
Culinary applications: *Gai-lan* can be stir-fried, boiled, or steamed.

Garlic

Garlic (*Allium sativum*) belongs to the same genus as onions and leeks in the amaryllis family (Amaryllidaceae). It grows from a fleshy bulb, made up of sections called cloves, which are the parts of the plant that are most commonly eaten. It is one of the most widely used ingredients to add taste and aroma to plant-based food.

Taste: Strong, sharp taste that becomes considerably mellower and sweeter after cooking; stimulates a *koku* sensation elicited by certain tripeptides.
Aroma: Pungent aroma due to sulfur-containing compounds, including allicin, released when the cloves are crushed or chopped.
Texture: Crisp when raw, becomes soft when cooked, and creamy when caramelised.
Culinary applications: Raw garlic is eaten crushed or finely chopped in salads and dressings. It can be roasted, grilled, boiled, and caramelised and is widely used to add flavour to savoury dishes.

Ginger

Ginger (*Zingiber officinale*) belongs to the very large ginger family (Zingiberaceae), which includes turmeric (*Curcuma longa*) and cardamom (*Elettaria cardamomum*). The rhizomes, and in some cases, the shoots are used as seasonings and taste additives.

Taste: Burning, spicy taste derived from a number of aromatic oils, namely gingerol, zingerone, zingiberene, and shogaol.
Aroma: Citrus-like.
Texture: Crisp when fresh; becomes woody as it ages.
Culinary applications: Ginger is added to both savoury and sweet dishes. It can be grated, fried, crystalised, marinated, juiced, and made into a sweet and sour pickle as *tsukemono* (*gari*). Ginger pairs well with relatively bland ingredients such as avocado, cucumbers, and zucchini and is used in baked goods.

Horseradish

Horseradish (*Armoracia rusticana*) is a perennial belonging to the cabbage and mustard family (Brassicaceae). Both the taproots and the leaves are edible.

Taste: The root is pungent and spicy, but the sensation dissipates quickly. Leaves are sharp, bitter, and peppery.
Aroma: Sulfurous compounds, isothiocyanates, are produced enzymatically when the root is grated or otherwise damaged, resulting in a very sharp taste and irritating aroma, but this can be prevented by heating the horseradish. The leaves have a milder aroma.
Texture: Generally crisp and juicy; roots may be woody when very large or old. Young leaves are usually tender but may become tough as they age.
Culinary applications: The roots are used as a spice or taste additive or grated as a condiment, often to enhance fatty dishes. The leaves can be prepared like other leafy greens and also made into pesto.

Jerusalem artichokes

Jerusalem artichokes (*Helianthus tuberosus*), also known as sunchokes, belong to the very large aster family (Asteraceae) and are related to sunflowers, artichokes, and lettuce, among other plants. The tubers, which resemble ginger roots, derive their energy from stored-up inulin. They are uneven, knobby, and difficult to peel. As they tend to turn brown when peeled or cut, they should be immersed in water with lemon juice or vinegar unless they are to be cooked right away.

Taste: Sweet, earthy, and a little bitter.
Aroma: Nutty.
Texture: Crisp, firm, juicy, and crunchy when raw; become mealy and soft when boiled.
Culinary applications: The tubers can be eaten raw in salads, but are usually baked, cooked, or puréed in soups. They can also be pickled as *tsukemono*.

Kale

Common kale (*Brassica oleracea* convar. *acephala*) is an annual plant belonging to the cabbage and mustard family (Brassicaceae). It does not form heads but has loose leaves with a curly edge that range in colour from light to dark green, as well as violet-green.

Taste: Slightly bitter and peppery, with typical sulfurous cabbage taste that is more intense after the leaves have been cooked.
Aroma: Earthy and grassy.
Texture: Tender and crunchy when raw; becomes softer when cooked.
Culinary applications: Kales are the most widely eaten types of cabbage and are used raw in salads, sautéed, stewed, and added to soups.

Kalettes

Kalettes (*Brassica oleracea* var. *viridis* × *gemmifera*), also known as flower sprouts, is the trade-marked name for kale sprouts, a relatively recent hybrid between kale and Brussels sprouts, both members of the cabbage and mustard family (Brassicaceae). They grow along the stalks like Brussels sprouts, but their small frilly heads are more open and grey-green and slightly purple in colour.

Taste: Mild, somewhat sweet, and slightly nutty.
Aroma: Mild sulfurous aromas typical of the cabbage family.
Texture: Crisp, but more tender than kale.
Culinary applications: Kalettes can be eaten raw, microwaved, roasted, and sautéed, or simmered in *dashi*.

Kohlrabis

Kohlrabis (*Brassica oleracea* var. *gongylodes*) belong to the cabbage and mustard family (Brassicaceae) but are distinct from rutabagas (*Brassica napus* ssp. *rapifera*). The swollen stems are eaten as vegetables and the leaves are treated as leafy greens. The skin on small, young ones can be eaten, but on older ones the skin can be tough and fibrous and should be peeled away. There are varieties with both green and purple skin.

Taste: The stem has a somewhat mild cabbage taste; the leaves are more bitter.
Aroma: Typical sulfurous cabbage aromas.
Texture: Crisp and crunchy when newly harvested; can become woody with age.
Culinary applications: Small kohlrabis can be eaten raw in salads and coleslaw. Mature ones are simmered or pickled, for example, as *tsukemono*.

Lacinato kale

Lacinato kale (*Brassica oleracea* var. *palmifolia*), also called Tuscany or dinosaur kale, belongs to the cabbage and mustard family (Brassicaceae). This leafy annual has very stiff, firm leaves that are dark bluish-green, almost black, and have a bumpy appearance.

Taste: Earthy, nutty; sweeter and milder than common kale.
Aroma: Earthy and herbal, but becomes sulfurous when wilting.
Texture: Tender, crisp.
Culinary applications: Lacinato is the best of the kales for using raw in salads, and is also used in stews and soups, especially minestrone.

Leeks

Leeks (*Allium ampeloprasum*) belong to the onion genus of the amaryllis family (Amaryllidaceae). Unlike the common onions that are bulbous, leeks are made up of a long cylinder of leaf sheaths. The strong onion taste develops in the same way as it does in garlic.

Taste: Sweet and somewhat meaty when cooked; sharp and strong when raw.
Aroma: Cabbage-like, slightly sulfurous aroma.
Texture: Crunchy when raw; smooth and slippery when cooked.
Culinary applications: Leeks can be boiled, steamed, toasted, grilled, or charred.

Lentils

Lentils (*Lens culinaris* or *Lens esculenta*) are legumes belonging to the pea family (Fabaceae). They are usually sold dried, identified by colour, for example, brown, red, and green. Small grey-green Puy lentils (*Lens esculenta puyensis*) are especially popular. The large black gram lentils (*Vigna mungo*), which are closely related to mung beans (*Vigna radiata*), provide visual contrast in a green salad. Because lentils, like beans, contain lectin, a protein that can cause digestive problems if the lentils are eaten raw, they should always be fully cooked.

Taste: Mild earthy taste.
Aroma: Nutty and peppery.
Texture: Can become mealy when cooked.
Culinary applications: Lentils are used in soups, stews, and curries, boiled, and tossed into salads, baked in bread, or roasted to eat as a snack.

Melons

Melons are members of the gourd family (Cucurbitaceae), related to cucumbers and squash, among others. The distinction is usually made between sweet melons, also called muskmelons, (*Cucumis melo* ssp.) and watermelons (*Citrullus lanatus*), but both types have similar flavour profiles. The flesh is usually yellow, orange, red, or pale green.

Taste: Various degrees of sweetness.
Aroma: Musky, fruity, and floral.
Texture: Crisp, juicy, and firm.
Culinary applications: All melons are generally eaten raw, often as desserts, but unripe melons can be used as vegetables. Watermelons can be juiced, their rinds can be pickled or boiled, and their seeds dried and roasted.

Mushrooms

Mushrooms are not plants but belong to the fungus kingdom. There are many different varieties of edible mushrooms, for example, common mushrooms (*Agaricus bisporus*) usually called button mushrooms when white, cremini when brown, and Portobello when large and mature, field mushrooms (*Agaricus campestris*), chanterelles (*Cantharellus cibarius*), porcini (*Boletus edulis*), shiitake (*Lentinus edodes*), horn of plenty (*Craterellus cornucopioides*), oyster mushrooms (*Pleurotus ostreatus*), and puffballs (*Calvatia gigantea*). Many of the common ones are dried and must be rehydrated before use. See also truffles.

Taste: Sweet, bitter, nutty, and umami, especially *shiitake*.
Aroma: Earthy.
Texture: Crisp to spongy.
Culinary applications: Mushrooms are often added to soups, stews, and risotto. They can be fried, stewed, dehydrated, pickled, or marinated.

Onions

Onions (*Allium* spp.) are part of a genus with many varieties belonging to the amaryllis family (Amaryllidaceae). They are made up of a swollen bulb covered with layers of what are actually leaves at the base of the stem. The most common cooking onions (*Allium cepa* spp.) come in yellow, red, and white varieties and are generally of medium size. Spring onions are small, young onions harvested before the stem base has become swollen. Other well-known onion types are the smaller shallots (*Allium cepa* var. *aggregatum*) that often grow in pairs and scallions (*Allium fistulosum*), also called green onions, that have hollow leaves and do not develop bulbs. When prepared with acidic ingredients, red onions lose some of their colour. The strong taste of onions develops in the same way as it does in garlic.

Taste: Sweet, umami, sharp and strong when raw. Red onions are sweeter than yellow ones.
Aroma: Sweet when fresh; can be sulfurous when the cells are cut.
Texture: Crisp texture when raw; soft and slimy when cooked.
Culinary applications: Onions, especially the red ones and scallions, are added to salads. They are used as vegetables on their own or to add taste and can be boiled, fried, caramelised, and pickled or added to soups and stews.

Parsley roots

Parsley roots (*Petroselinum crispum* var. *tuberosum*) resemble parsnips and also belong to the umbellifer family (Apiaceae) but are smaller and have a more delicate texture. The edible leaves are like those of common curly parsley (*Petroselinum crispum*).

Taste: Strong, sharp, and a little sweet; similar to celeriac.
Aroma: Herbal aroma from terpenoids.
Texture: Crisp when raw; creamy and smooth when cooked.
Culinary applications: The roots can be eaten raw or prepared with heat in virtually any way. They are added to soups and stews, can be marinated or dehydrated, and toasted to make chips.

Parsnips

Parsnips (*Pastinaca sativa*) are root vegetables belonging to the umbellifer family (Apiaceae), along with carrots, parsley, celery, dill, and fennel. Parsnips have less starch than potatoes but more than carrots. They are less sweet than carrots but become sweeter after they have been exposed to frost because the starch is turned into sugar. They resemble parsley roots but are larger and coarser in texture. They oxidise quickly when peeled or cut and should be immersed in slightly acidic water if not intended for immediate use.

Taste: Sweet and slightly bitter, with faint notes of liquorice.
Aroma: Earthy, nutty aroma.
Texture: Crisp, but older ones become woody.
Culinary applications: Parsnips are rarely eaten raw. They can be prepared like other root vegetables but are best when roasted. Sometimes they are dehydrated and toasted to make chips.

Pears

Pears (*Pyrus* spp.), of which there are more than 3,000 known varieties, belong to the rose family (Rosaceae). Although pears are traditionally considered a sweet ingredient, green, unripe pears can be prepared as a vegetable.

Taste: Generally, slightly sour and only a little sweet when unripe.
Aroma: Depends on the variety, but generally fruity.
Texture: Crisp and juicy; very firm when unripe.
Culinary applications: Raw pears are added to salads. They can also be simmered, pickled, and juiced.

Peas

Peas (*Pisum sativum*) are legumes belonging to the pea family (Fabaceae), along with beans and lentils. They are actually fruits but are usually thought of as vegetables. Common garden peas are encased in pods and are normally shelled before use. Unripe and young ones can be eaten whole with their shell, as can the tender shoots and leaves of the plant. When dried they come in yellow and green varieties. Two popular pea varieties, snow peas and sugar snap peas, both classified as *Pisum sativum* var. 'Macrocarpon Group,' are always eaten whole. The former have flat pods with thin shells and the latter have slightly thicker, rounded shells.

Taste: Sweet and very rich in umami.
Aroma: Derived from aldehydes and similar to those of green bell peppers.
Texture: Garden peas are soft when freshly picked but become starchy over time. Shells of snow and sugar snap peas are crispy and crunchy.
Culinary applications: Garden peas can be eaten raw, boiled, fermented, and dehydrated. They are also puréed in hot and cold soups or to make a dip or hummus. Snow and sugar snap peas are eaten in salads and in stir-fries. Dried peas are rehydrated and prepared like lentils.

Potatoes

Potatoes (*Solanum tuberosum*) belong to the nightshade family (Solanaceae). There are many different varieties, and they are cultivated worldwide. Potatoes set large tubers that are very starchy and are an excellent energy source. Their tastes and aromas can be quite different depending on the variety and whether the potatoes are new, that is, immature tubers, or old ones that might have been harvested much earlier.

Taste: Often a little sweet with a bit of bitterness and a mild earthy taste due to pyrazine formation by soil microbes. Cooked new potatoes have a predominantly sweet taste, while old potatoes have a more intense taste that is due to a slow-acting enzyme that creates round, fruity, and flowery notes by breaking down lipids in the cell membranes. Potatoes that have turned green are very bitter due to the presence of alkaloids and it is unsafe to eat them. Unpeeled cooked potatoes, especially old ones, have umami taste substances that seep out into the cooking water.
Aroma: When raw, potatoes smell earthy, nutty, and starchy. Potatoes cooked with fats and stored in the refrigerator for several days develop a stale aroma and taste because their vitamin C content is no longer sufficient to prevent the fats from oxidising and becoming rancid.
Texture: Potatoes are often divided into three groups—mealy, firm, and waxy—according to their texture after they have been cooked. Mealy potatoes have a high starch content and easily overcook. They are suitable for baking, mashing, and making potato soup. Generally, some butter or other fat is added to the mash to ensure that it has a creamy consistency. Firm potatoes have less starch and have a wide variety of uses. Waxy potatoes have the least starch and become quite firm when cooked. They are, therefore, used to make potato salad or cut up to be roasted or fried. If waxy potatoes are puréed, they can become rubbery and it is difficult to incorporate any fat. Some potatoes have a thin peel that is very edible.
Culinary applications: Potatoes cannot be eaten raw, but otherwise can be prepared in virtually all ways. Baking, frying, deep-frying, or roasting potatoes leads to the formation of Maillard compounds with a characteristic malty, sweet taste and aroma.

Radicchio

Radicchio (*Cichorium intybus* var. *foliosum*) is a leaf vegetable belonging to the aster family (Asteraceae). It grows in the form of small, compact heads that have a deep red colour.

Taste: Bitter, spicy.
Aroma: Mild.
Texture: Crisp.
Culinary applications: Radicchio can be made less bitter by soaking cut leaves in icy water. It is eaten raw in salads, partly to add visual appeal. It can also be grilled and pickled in a sweet and sour marinade.

Radishes

Radishes (*Raphanus* spp.) belong to the cabbage and mustard family (Brassicaceae). Their skin can be white, pink, red, purple, yellow, green, or black, but the inside is usually white. The most common are the red *Raphanus sativus* var. *sativus* that have round and elongated shapes. Black radishes (*Raphanus sativus* var. *niger*) are larger and have a tough black skin and white flesh. When sliced crosswise, they contribute an important visual aspect to a dish.

Taste: Strong, peppery, stinging, and sharp; generally, most noticeable in the outer layer.
Aroma: Pungent aroma due to concentration of isothiocyanates and thiocyanates.
Texture: Crisp and crunchy when freshly harvested; can be woody when old.
Culinary applications: Radishes are eaten raw in salads or served as crudités and with fresh cheeses. They can also be roasted, marinated, pickled, and made into *tsukemono*.

Red Russian kale

Red Russian kale (*Brassica napus* var. *pabularia*) belongs to the cabbage and mustard family (Brassicaceae). It is a leafy annual with bright burgundy stems and flat, serrated, purple-tinted leaves.

Taste: Mild, nutty, slightly sweet, and earthy.
Aroma: Earthy, with slight sulfurous overtones when older.
Texture: Tender but firm, crisp leaves and fibrous, tough stems.
Culinary applications: Like other kales, the leaves can be eaten raw, sautéed, stewed, and added to soups. The stems are edible and can be roasted.

Rhubarb

Rhubarb plants (*Rheum* spp.) are perennials belonging to the knotweed family (Polygonaceae). The fleshy, edible stalks are actually vegetables, although we think of them as fruits to be used in sweet dishes.

Taste: Sour taste due to several different organic acids, such as oxalic acid, citric acid, and malic acid. Also astringent.
Aroma: Tart, fresh, and fruity.
Texture: Crisp, meaty, and juicy, but old ones are woody.
Culinary applications: Rhubarb stems are boiled to make compotes and can be pickled. They are also used in baked goods and ice cream.

Rutabagas

Rutabagas (*Brassica napus* ssp. *rapifera*), also known as swedes, are a hybrid between ordinary cabbage (*Brassica oleracea*) and turnips (*Brassica rapa* ssp. *rapa*) and should not be confused with kohlrabi. Rutabagas are coarser and can be woody.

Taste: Slightly bitter, sweet, and earthy; similar to potatoes when cooked.
Aroma: Characteristic cabbage aromas.
Texture: Crisp and very hard.
Culinary applications: Rutabagas can be prepared like other root vegetables, pickled, and smoked.

Salad greens

There is a large group of green plants with edible leaves and stems, broadly known as sweet lettuces (*Lactuca sativa* convar.) that belong to the aster family (Asteraceae). They do not have significant nutritional value and apart from their mild, sweet taste, crisp textures, and attractive green and red colours, they add little to a meal. Some of the more popular varieties are: head lettuce (*Lactuca sativa* var. *capita*), which has thin, soft leaves and a mild taste; iceberg lettuce (*Lactuca sativa* var. *capitata crispum*) with somewhat thicker leaves and very juicy, crisp leaves; romaine (*Lactuca sativa* var. *longifolia*), which has long, firm leaves, but only the inner ones are eaten; leaf lettuce (*Lactuca sativa* var. *crispa*) with crimped loose leaves that are green or tinged with burgundy.

Taste: Mild and sweet.
Aroma: Herbal and grassy.
Texture: Juicy and crisp.
Culinary applications: Used in cold salads and as garnishes.

Seaweeds

Seaweeds are marine macroalgae, not plants. There are about 12,000 different species, belonging to very diverse biological groups. From a culinary perspective, but not a scientific one, they are divided into green, red, and brown ones. Apart from a few species, they are all edible. Their tastes, aromas, and textures reflect the great variations among them and are very dependent on how they are prepared. Seaweeds are probably more versatile than any other ingredient used in green cuisine. They are eaten raw on their own or combined with a vast number of other cold or warm ingredients. They can be boiled, baked, toasted, puréed, dehydrated, granulated, or deep-fried, as well as incorporated into snacks, appetisers, entrées, and desserts. Seaweeds feature both in everyday meals and in gastronomic specialties that take advantage of their special qualities as a challenging and interesting raw ingredient. An easy way to start to cook with seaweeds is to use them as a salt substitute in bread, as a spice, or simply to toss them in a green salad. *Konbu* (*Saccharina japonica*) and dulse (*Palmaria palmata*) are especially good sources of umami. Among the commonly eaten species are: the green seaweeds, sea lettuce (*Ulva lactuca*) and green string lettuce (*Ulva linza*); the red seaweeds, dulse (*Palmaria palmata*) and laver also called *nori* (*Porphyra/Pyropia* spp.); the brown seaweeds, *konbu* (*Saccharina japonica*), *wakame* (*Undaria pinnatifida*), winged kelp (*Alaria esculenta*), sugar kelp (*Saccharina latissima*), and bladderwrack (*Fucus vesiculosus*).

Spinach

Spinach (*Spinacia oleracea*) belongs to the amaranth family (Amaranthaceae), as do beets. Both the leaves and stalks are edible. It is one of the most important sources of folate (vitamin B_9). The astringent mouthfeel of spinach is due to its oxalic acid content.

Taste: Mild and slightly sweet when raw; somewhat bitter when cooked.
Aroma: Fruity, nutty, and herbal.
Texture: Slightly astringent, meaty, and soft.
Culinary applications: Spinach is eaten raw in salads and blended into smoothies or boiled, wilted, sautéed, and steamed.

Squash

Squash are fruits from plants in the gourd family (Cucurbitaceae) that grow along the ground and are usually prepared as vegetables. They tend to be quite large and have a very hard peel and somewhat soft flesh in various shades of yellow and orange. Some well-known ones are dark green acorn squash (*Cucurbita pepo* var. *turbinata*), reddish-orange Hokkaido squash (*Cucurbita maxima* Duchesne), bell-shaped yellow butternut squash (*Cucurbita moschata*), and mottled green kabocha squash (*Cucurbita maxima* var. *kabocha*). They are normally peeled but the peel of some varieties can be softened by roasting or baking. Seeds from some varieties can be toasted and used as Flavour Accelerators, for example, simmered in soy sauce.

Taste: Slightly sweet, buttery, and nutty, depending on variety.
Aroma: Fresh and citrusy.
Texture: Soft and creamy when baked or cooked.
Culinary applications: Squash are not eaten raw. They can be boiled, simmered, grilled, baked, pickled, and puréed for soups and are also used in baked goods.

Strawberries

Strawberries (*Fragaria* spp.) belong to the rose family (Rosaceae) and there are more than twenty different species in the genus. Botanically speaking they are considered to be 'accessory fruits' and are not true berries. The edible flesh consists of the swollen flower receptacle that is studded with tiny seeds. When they are unripe and green they can be used as vegetables as their taste, aroma, and texture are very different from those of ripe red ones.

Taste: Slightly bitter and sour when unripe, becoming sweet as they ripen.
Aroma: Grassy when unripe; fruity and fragrant when ripe due to aromatic compounds.
Texture: Very firm and somewhat dry when unripe; juicy and soft when ripe.
Culinary applications: Unripe ones are added to salads or pickled. Ripe ones are generally eaten in deserts.

Sweet potatoes

Sweet potatoes (*Ipomoea batatas*) belong to the morning glory family (Convolvulaceae). Of its more than 1,600 species, they are the only food crop of any major importance. They are only distantly related to ordinary potatoes and should not be confused with yams (*Dioscorea* spp.). The nutritional content of sweet potatoes is similar to that of potatoes, but they have more proteins, sugar, and dietary fibre. In addition, they are also a source of vitamin A, but have less vitamin C than potatoes. When heated (57–75°C or 135–165°F), sweet potatoes become very sweet—hence the name—and almost syrupy because they have an enzyme that breaks starch down to maltose. That is why baking them slowly results in a sweeter taste than boiling them at high heat. Those with a red skin become more tender and juicier when cooked, but those with a yellowish skin become mealier and drier.

Taste: Nutty and sweet, especially when baked at low temperatures.
Aroma: Floral and fruity.
Texture: Soft, creamy, and possibly mealy when cooked.
Culinary applications: Sweet potatoes can be boiled, baked, puréed, and mashed.

Swiss chard

Swiss chard (*Beta vulgaris* convar. *cicla*), which is a variant of the wild sea beet (*Beta vulgaris*), belongs to the amaranth family (Amaranthaceae). Both the green leaves and stems are edible and there are varieties that have white stems (silver beets), as well as yellow and red ones.

Taste: Little taste, somewhat bland, and slightly bitter.
Aroma: Fresh herbal aroma.
Texture: Juicy, crisp leaves; the stems can be somewhat woody.
Culinary applications: Swiss chard is eaten raw in salads. It can be steamed, sautéed, and substituted for spinach in recipes.

Tomatoes

Tomatoes (*Solanum lycopersicum*) belong to the nightshade family (Solanaceae). There are more than 7,000 different varieties and hybrids, including garden tomatoes, cherry tomatoes, grape tomatoes, plum tomatoes, and beefsteak tomatoes. While most are red, there are also yellow, orange, brown, and green cultivars. Although they are true fruits, they are treated as vegetables and globally are the most widely used ingredient in that category. Sun-ripened tomatoes are a wonderful source of umami, which is more intense in the pulpy interior than in the outer flesh. In sun-dried tomatoes it is even more concentrated.

Taste: Sweet, tart, tangy, and rich in umami.
Aroma: Smoky and fruity due to a long series of volatile compounds, especially hexanal, beta-ionone, beta-damascenone, 1-penten-3-one, and 3-methylbutanal. When heated, some sulfurous aroma substances are formed.
Texture: Firm and juicy; can be somewhat mealy.
Culinary applications: Tomatoes are eaten raw as a snack and in salads. They can be prepared in virtually any way for all types of dishes and are sometimes dehydrated or pickled.

Truffles

Truffles (*Tuber* spp.), like mushrooms, are fungi and belong to the Tuberaceae family, but are really in a category by themselves. They grow entirely underground, under very special conditions in symbiosis with the roots of a number of tree species. When truffles are mature and ready to disperse their spores, they give off a potent musky odour. Their taste is earthy and a little sulfuric (from dimethyl sulfide) with notes of sweet fruits. Because they release this heady aroma when heated, they should be added to a dish just before serving. Summer truffles (*Tuber melanosporum*) have a milder taste and aroma than winter ones (*Tuber aestivum*), which have a deep, strong flavour. White truffles (*Tuber magnatum*) have a fuller flavour than the black ones and their taste is a little sharper and more bitter, but at the same time more subtle and aromatic. As a result, they are highly prized and are more expensive.

Taste: Umami, earthy, sweet; can be bitter.
Aroma: Musky and sulfuric odours.
Texture: Firm, slightly crunchy.
Culinary applications: Truffles have many uses with vegetable dishes and pair particularly well with white asparagus and potatoes and with pasta dishes. Small quantities are sometimes shaved over a warm dish as it is being served. Truffles are also incorporated into other foods such as olive oil, vinegar, salt, and cheeses.

Turnips

Turnips (*Brassica rapa* ssp. *rapa*) belong to the cabbage and mustard family (Brassicaceae) and are similar to radishes. The swollen round storage roots have a white skin, but the part of the turnip that sits above the ground may be purple, red, or greenish. The leaves are also edible, especially those from young ones.

Taste: Characteristic, but quite mild, cabbage taste.
Aroma: Spicy, pungent, and a little like horseradish.
Texture: Meaty, crisp, and crunchy when raw; older ones can be woody; may become mushy if overcooked.
Culinary applications: Turnips are eaten raw in salads and can be boiled, steamed, glazed, and pickled to make *tsukemono*.

Yams

Yams (*Dioscorea* spp.) are the cylindrical tubers of a tropical plant related to grasses. Yams come in many different species and must not be confused with sweet potatoes. Along with potatoes, cassava, and sweet potatoes, yams are a staple, starchy food for a large part of the world population. The tuber has a white or yellow to purple, soft inside and a tough, dark skin that can be softened by cooking. Yams have a very high content of potassium and a glycaemic index similar to that of potatoes. Some African species must be boiled to be safe for consumption. Yams easily absorb the taste and aroma of other ingredients in a dish.

Taste: Similar to potatoes and not as sweet as sweet potatoes.
Aroma: Mild and very faint; nutty when roasted and grilled.
Texture: Similar to potatoes; when cooked can have a slimy and frothy appearance from saponins.
Culinary applications: Cooked or roasted and used in the same ways as potatoes and sweet potatoes, particularly cooked, baked, fried, and roasted and eaten without the skin.

Zucchinis

Zucchinis (*Cucurbita pepo*), also called courgettes, are summer marrows belonging to the gourd family (Cucurbitaceae). Because zucchinis are actually immature marrows, their skins have not hardened, and they can be eaten raw with their peels. They are usually cylindrical, although some are large and round, and have dark green, pale green, yellow, and orange skins and whitish insides.

Tastes: Bland, slightly sweet.
Aromas: Mild and grassy. The flowers have a musky, spicy aroma with herbal and floral notes.
Textures: Crisp and watery when raw; very soft and creamy when cooked.
Culinary applications: Zucchinis can be eaten raw or simmered, roasted, and grilled. They are often added to stews and sauces, especially ratatouille. When grated, can be baked in muffins and cakes. The flowers can be deep-fried as is or stuffed with meat and cheese.

THE FLAVOUR ACCELERATORS

adding umami and texture

The Flavour Accelerators are home-made condiments and off-the-shelf products that are especially useful for adding taste, aroma, and texture to green cuisine. Apart from a few traditional spices such as pepper and powdered chilli, these are not spices in the conventional sense. They are ingredients that are particularly well suited for enhancing the flavour of vegetables. These items are mostly ones that have a good shelf-life, that can be purchased or prepared in advance, and that can be kept on hand in the pantry, the refrigerator, or possibly the freezer. We have not included grains, eggs, and dairy products apart from cheese on this list. In quite a few cases there are examples of the recipes in which they have been used. The arrows (») cross-reference other entries in the list.

Because their impact is often based on a complex interaction of tastes, aromas, and *koku* attributes, we have characterised this as their flavour profile.

Aïoli

Aïoli is a type of »mayonnaise heavily flavoured with »garlic. Certain tripeptides in the garlic contribute to a *koku* sensation (*koku* attribute). It is commonly served as a condiment for seafood and can also be used as a dip for crudités.

Aïoli

2 egg yolks	1 clove of garlic
Pinch of salt	A few drops of lemon juice
1 tsp Dijon mustard	Salt and pepper
1½ dl (⅗ c) grapeseed oil	

1. Whisk the egg yolks with the salt and mustard until light coloured and slightly set.
2. Add the oil one drop at a time to start, while continuing to whisk. Adding too much oil at one time will cause the mayonnaise to separate.
3. Continue whisking until all the oil is added and the mixture has the consistency of a mayonnaise.
4. Put the garlic clove through a press and whisk it into the mayonnaise. Season to taste with salt, pepper, and lemon juice.
5. Refrigerate if not using right away. The *aïoli* will keep for weeks.

Flavour profile: Salty and slightly sour, strong garlic taste; peptides in garlic contribute to a *koku* sensation; aroma from sulfur compounds.
Texture: Creamy.

Anchovies, anchovy paste, anchovy sauce

Salted anchovies (*Engraulis encrasicolus*) and anchovy paste are prominent ingredients in many Mediterranean dishes. The considerable glutamate and inosinate contents of the fish contribute a very strong umami synergy. Salted anchovies are also found in »ketchup and »Worcestershire sauce. Anchovies are available in cans and anchovy paste is sold in tubes that need to be stored in the refrigerator. They are used as taste additives in vinaigrettes, dressings, marinades, pizza toppings, and »pesto. Anchovy sauce (*bagna càuda*) is used as a dip for a large variety of raw vegetables and on cooked ones.

Flavour profile: Umami, salty, and sharp tastes; smell of the sea and pungent fish aromas.
Texture: Fillets are firm and succulent, the paste is creamy.
Recipes: Grilled onions, p. 238. Mushroom pâté, p. 262. Carrots in zucchini sauce, p. 228.

Anchovy sauce (*bagna càuda*)

10 cloves of garlic	100 g (3 ½ oz) butter
4–5 dl (1 ⅔–2 ⅛ c) milk for blanching	2 dl (⅘ c) olive oil
200 g (7 oz) salted anchovy fillets	2–3 tbsp breadcrumbs, preferably *panko*
	Freshly ground white pepper

1. Peel the garlic cloves and blanch them in a little milk and water three times, rinsing them under cold running water each time.
2. Bring the remaining milk to a boil (with no added water) and blanch and rinse the garlic cloves one final time. Keep the milk.
3. Soak the anchovy fillets for 10 minutes in cold water.
4. Blend the garlic, anchovy fillets, and butter while adding the olive oil.
5. Add in the breadcrumbs and season with white pepper.
6. If necessary, add more breadcrumbs and the warm reserved garlic milk to achieve a consistency that is suitable for how the sauce is to be used.

Flavour profile: Salty, strong garlic taste, umami, *koku* sensation; fish smell and aroma from sulfur compounds in garlic.
Texture: Creamy, finely granulated.
Recipe: Cone cabbage with asparagus, p. 226.

Bacon

Bacon is salted and cured meat from land animals, typically pork, but also from beef, poultry, and game. It can also be cold smoked. In the case of pork, both belly and back cuts are chosen, depending on how fatty the finished bacon is to be. They can be cured by either wet or dry salting. Nitrite is often added as a preservative and to produce the characteristic pink colour of the meat parts. In the case of wet salting, different ingredients, such as maple syrup, sugar, or tomatoes may be added to produce distinct tastes. Smoked bacon reflects the aroma from the type of wood used in the process. Bacon can be fried on its own or cooked as part of a dish. It should never be eaten raw.

Flavour profile: Salty, possibly sweet, meaty; may have smoky nuances depending on type of wood used and aromas from seasonings.
Texture: Firm and chewy, fatty but crisp when fried at high heat.
Recipe: Pears, beans, potatoes, and smoked port belly, p. 248.

Balsamic vinegar

Balsamic vinegar is a special type of »vinegar made from a grape juice reduction that is aged in wooden barrels for at least one year and often for decades. Using what is known as the Solera method, vinegar from younger batches is blended progressively with older ones over a period of years to produce a delicious, sweet, thick liquid with a unique and complex character.

Flavour profile: Very complex umami, sweet, and sour tastes; aromatics from the various types of wood in the aeging barrels.
Texture: Viscous liquid, sticky.
Recipe: Grilled onions, p. 238.

Black/caramelised garlic

Even though black garlic is sometimes described as 'fermented,' it is actually prepared by storing fresh »garlic at 60–80°C (140–175° F) and 70–80% humidity for 4–6 weeks. During this time Maillard reactions give rise to dark, flavourful compounds. The garlic turns completely black, and the sharp garlic taste almost disappears and it acquires complex, milder tastes. At the same time, the cloves lose their crispness and become soft and a little waxy. Onions and leeks can be prepared in the same way. Black garlic can be chopped and sprinkled on green dishes and salads or mashed and mixed into a dressing or sauce.

Flavour profile: Umami, mild, slightly acidic, sweet, pleasantly rounded taste with notes of »balsamic vinegar and »tamarind; exhibits *koku* sensation; mild garlic aroma.
Texture: Soft, creamy, a little waxy, and a bit sticky.

Blue-veined cheeses

Blue cheeses are inoculated with a culture of either *Penecillium roqueforti* or *Penicillium glaucum*. Among the best-known ones are French Roquefort and Bleu de Bresse, English Stilton, Italian Gorgonzola, and Danish Danablu. Some of these cheese types are semi-soft and creamy, while others are firm, dry, and crumbly. As the cheeses ripen, enzymes break down milk proteins to produce large quantities of free amino acids, especially glutamates, and consequently a great deal of umami. Because of their strong, complex taste profile, blue-veined cheeses are not served only as part of a cheese plate but are also crumbled into salads and added to dressings, dips, and sauces.

Flavour profile: Umami, salty, sour, *koku* sensation; pungent odour with traces of ammonia.
Texture: Creamy, dry, crumbly.
Recipe: Polenta fritters with blue cheese dip, p. 256.

Botargo

Botargo is dried fish roe, usually from tuna, cod, or grey mullet, which is considered a great delicacy in Spain (*botargo*) and Italy (*bottarga*). The roe pouches are removed whole from the fish, then cured with sea salt for a few weeks, and finally hung up to air dry for a month. The salt draws liquid out of the roe, making it firm and hard. Like other fish roes, *botargo* has a rich umami taste from glutamate and inosinate. It can be grated and sprinkled on vegetable dishes.

Flavour profile: Umami, salty; mild briny smell of the sea.
Texture: Firm, waxy.

Bouillon

A bouillon is an aqueous extract, made by straining the liquid in which some combination of meat and/or bones from beef, veal, and pork, poultry, fish, or shellfish, as well as vegetables and mushrooms have been cooked in water. This results in a liquid with optimal umami synergy that can be used as the basis for many soups and sauces and to prepare vegetables. Commercially prepared bouillon is available in cans and packages or in dried form in cubes and powders. Reducing 1 litre (4 ¼ cups) of bouillon to 1 decilitre (²/₅ cup) results in a thick liquid that can be used in dressings or sprinkled sparingly on a salad. 'Stock' is a related term that generally refers to an extract that includes bones, resulting in a thicker liquid due to the collagen content of the bones. Similarly, 'broth' usually denotes an extract that is made with raw meat, other ingredients, and no bones. Japanese »*dashi*, an essential ingredient in many

The Flavour Accelerators

Japanese dishes, is made as an aqueous extract of a seaweed, »*konbu* (*Saccharina japonica*), and certain fish flakes, »*katsuobushi*, or dry *shiitake* in water.

Flavour profile: Umami; other tastes and aromas depend on specific ingredients and any seasonings that might have been added.
Texture: Liquid, consistency depends on fat and collagen content.
Recipes: Onion soup—the real thing! p. 190. Juicy aubergines, p. 212.

Basic vegetable bouillon

3 l (12 ¾ c) water or water from cooking potatoes	5 g (1 tsp) chilli powder, medium hot
1 kg (2 ⅕ lb) vegetables (for example, celery, leeks, onions, carrots)	3 cloves of garlic
	6 laurel leaves
250 g (½ lb) tomatoes	1 sprig of thyme
125 g (¼ lb) mushrooms	5 g (⅕ oz) *konbu*
20 peppercorns	10 g (2 tsp) sea salt (if using cooking water, this amount can be reduced)

1. Rinse all vegetables, cut them up into large pieces, and put them in a pot with the rest of the ingredients.
2. Allow the bouillon to simmer covered for 30–40 minutes until the vegetables are very soft. Turn off the heat and leave the vegetables in the water for about an hour. Add the *konbu* at about the halfway point. Strain the bouillon.
3. The bouillon can be frozen if it is not needed right away.

Bread

Leftover bread can be cubed and toasted or made into crumbs and stored in an air-tight container for a long time. They should be heated on a warm skillet before use, for example, in the form of croutons for use on a salad. Due to their particular airy structure, »*panko* breadcrumbs absorb less fat or oil when fried, which results in a lighter, crisper crust, for example, for tempura. »*Rouille* is a sauce thickened with breadcrumbs or the crusts from old bread.

Flavour profile: Sweet, depending on type of bread used; taste and possibly aroma from Maillard compounds.
Texture: Crisp, crunchy.
Recipe: Grilled onions, p. 238.

Capers and caper berries

Capers are made from the small flower buds of the caper bush (*Capparis spinosa*), while caper berries are the fruits, which are larger and picked while immature with their stems attached. Both buds and berries are pickled with salt and vinegar and used in sauces, salads, and other vegetable dishes. They can be deep-fried and become very crisp. There is some umami in capers and their slightly sharp taste is due to methylisocyanate, a volatile compound also found in »mustard.

Flavour profile: Salty, sour, sharp; mildly briny, vinegary smell.
Texture: Capers are juicy and succulent; caper berries are firm with crunchy seeds.
Recipes: Green asparagus, p. 220. Succulent leeks, p. 240.

Cephalopods

Cephalopods (octopuses, squid, and cuttlefish) are a somewhat overlooked resource in some food cultures. The flesh from their arms, tentacles, and mantle is firm and has a very smooth texture that can easily be substituted for many types of meat from terrestrial animals. It is has very little fat but is rich in proteins and is characterised by the considerable content of nutrients that are important sources of sweetness and umami. Achieving the right mouthfeel of prepared cephalopods can be a challenge, but armed with a little knowledge it is possible to reach the desired texture, be it tender, chewy, firm, or succulent. Even in small quantities, cephalopods are able to introduce taste, especially umami, and texture with a bit of bite to vegetable dishes and salad. Dried cephalopods are especially useful for adding texture. Marinated and dehydrated mantles and fins from large squid, for example, *Loligo forbesii*, can be grated and sprinkled on top of a dish, a little like »*botargo* and »Parmesan cheese. Strips of squid or cuttlefish mantle meat prepared as a confit are an excellent complement for a green salad. Smoking adds extra taste and aroma to dehydrated cephalopods. Thin slices of the arms of cooked and smoked octopuses (*Octopus vulgaris*) can be tossed in a salad or added to a soup.

Flavour profile: Umami, salty; mild marine aroma, can have a smoky smell depending on how they are prepared.
Texture: Firm and chewy, tough, creamy.

Charred onions

Charring onions (or other species in the *Allium* family, such as leeks) gives rise to pyrolysis products and Maillard compounds. Roasted onions have a milder taste than charred onions. If they are roasted with sugar, they can be caramelised and are excellent for flavouring and colouring dark bouillons.

Flavour profile: Burnt, possibly caramelly, *koku* sensation; taste and aroma from Maillard reactions.
Texture: Soft with crisp edges.

Chilli peppers and powders

On a worldwide basis, chilli is the most used spice next to salt and followed by pepper. Despite their name, chilli peppers are not actually »peppers, but are fruits belonging to the *Capsicum* genus in the nightshade family (Solanaceae). The taste experience of chilli peppers is strong, irritating, and burning (chemesthetic action, trigeminal sensation) caused by compounds called capsaicinoids that interact with the mucosal membranes in the mouth. There are many species and hybrids of *Capsicum* and the intensity of their hot taste varies greatly. Habñeros are among the hottest, serranos and jalapeños are of medium strength, and poblanos and Anaheim the mildest. Chilli spice is available in the form of powders and flakes, which have a long shelf-life. These are the 'go-to' spices for adding heat to virtually any dish.

Flavour profile: Strong, burning; aromas vary by type from mild and grassy to floral and fruity.
Texture: Crisp and juicy when raw; granular or powdery when dehydrated to make a spice.
Recipe: Chilli watermelon, p. 270.

Citrus fruits

There is a large number of species, hybrids, and cultivars in the *Citrus* genus. Common ones include lemon, lime, orange, pomelo, and grapefruit, with bergamot and »*yuzu* being less well-known. They are all characterised by acidity from citric acid and aroma from limonene but are differentiated by their sweetness and aroma compounds. *Yuzu*, which is a key ingredient in »*ponzu*, goes especially well with vegetable dishes. The rind of citrus fruits is particularly rich in aromatic oils and can be used as a flavouring agent to impart both acidity and aroma. Citrus fruits can be salt-pickled and fermented for use as condiments. They are eaten raw or in salads and can be incorporated into cooked dishes. Their juices are used to add an aromatic touch of acidity and are ingredients in many dressings and marinades. Dried bits of rind are found in spice mixtures, for example, »*shichimi*.

Flavour profile: Range from sweet to sour, possibly bitter; aroma from limonene.
Texture: Viscous when pulped; rind is soft unless dried.
Recipe: Salt-pickled lemons, p. 174.

Citrus dressing

1 dl (²⁄₅ c) freshly squeezed orange juice	1 tbsp Worcestershire sauce
	3 tbsp tomato ketchup
½ dl (⅕ c) freshly squeezed lime juice	½ tsp honey
	¼ tsp ground pepper
¼ dl (1 ²⁄₃ tbsp) freshly squeezed lemon juice	2 spring onions

1. Mix together all the ingredients except the spring onions.
2. Chop up the spring onions very finely and mix into the dressing.

The Flavour Accelerators

Dashi

Dashi (from Japanese for 'cooked extract') is an aqueous extract of a »seaweed, *konbu*, (*Saccharina japonica*), and bonito fish flakes, »*katsuobushi*. *Ichiban dashi* and *niban dashi* refer to the first and second extracts, respectively. The first extract has the mildest taste and the most delicate aroma. *Konbu-dashi* is made using only the seaweed. *Shōjin dashi* is a vegetarian/vegan *dashi* in which »shiitake are substituted for *katsuobushi*. *Dashi* is the ubiquitous source of umami in Japanese cuisine. Due to the interaction between glutamate from the seaweed and inosinate from the fish flakes or guanylate from the mushrooms, the extract displays perfect umami synergy. Freshly made *dashi* will keep in the refrigerator for several days. It can also be made from commercial powder products (*hon-dashi*) that are easy to dissolve in water. *Dashi* is an important ingredient in Japanese sauces such as »*ponzu* and »*sanbaizu*. It is also used as a marinade and for steaming vegetables.

Dashi

500 ml (2 ⅛ c) water, cold
5 g (⅕ oz) *konbu*
12 g (2 ½ tsp) *katsuobushi* flakes

1. Soak the *konbu* in the water in a pot for 30 minutes.
2. Heat to ca. 60°C (140°F) for 30 minutes.
3. Remove the seaweed and heat the liquid to 60–70°C (140–160°F).
4. Sprinkle the fish flakes into the water.
5. When the flakes have sunk to the bottom, skim off any foam that may have formed. Pass through a sieve to remove the fish flakes.
6. The remaining liquid is *dashi*.

Flavour profile: Umami, slightly smoky; smell of the sea.
Texture: Liquid.
Recipes: *Miso* soup with seaweed, p. 188. Simmered *daikon*, p. 234.

Fish flakes

Fish flakes are small flakes of dried, typically lean, fish that is less susceptible to becoming rancid from the fat content. They are easy to sprinkle over salads or use in soups and vegetable dishes, and even in small quantities add taste, especially umami, and contribute texture. See also »*katsuobushi*.

Flavour profile: Umami, salty; mild fish smell.
Texture: Dry, possibly chewy.

Fish sauce

Fermented sauces made with fresh or dehydrated fish, shellfish, and molluscs are common throughout Southeast Asia. The sauces are produced by salting and fermenting the marine animals, using them whole or only parts of them, such as blood and entrails. »Anchovies are an ingredient in many of them. The fermentation process depends on the animals' own intestinal enzymes, which are often augmented by adding the hepatopancreas (liver) from squid. Large amounts of free amino acids, especially glutamates, and peptides are formed, resulting in umami and *koku* sensation. Most fish sauces contain large quantities of salt. »*Garum* is a fish sauce that originated in the Mediterranean areas of the ancient world. Unlike anchovy paste, fish sauces have no free nucleotides. Fish sauces are used in marinades and dressings and can be added in small amounts to add umami and *koku* sensation to stews and soups.

Flavour profile: Umami, salty; long periods of fermentation result in nut and cheese tastes, *koku* sensation; pungent fish odour.
Texture: Liquid, slightly viscous.
Recipes: *Koji*-marinated broccolini with chilli sauce, p. 178. Kimchi, p. 179. Tomatoes in tomatoes, with mouthfeel, p. 268.

Fruit vinegar marinade

There are many uses for a marinade of this type in vegetable dishes. The characteristic aromas of the vinegar serve to modify the pungent effect of the mustard seeds.

Fruit vinegar marinade

4 tsp yellow or black mustard seeds	1 dl (2/5 c) fruit vinegar, for example, plum (*umesu*), raspberry, black currant, or pomegranate vinegar
½ dl (1/5 c) water	
½ tsp salt	3 tbsp honey

1. Put the mustard seeds and water together in a pot, bring them to a boil, and then drain off the water.
2. Add the vinegar, salt, and honey and cook the mustard seeds for 2 minutes. Set aside until used.

Flavour profile: Sour, sweet, strong; aroma varies depending on the particular fruit vinegar.
Texture: Liquid with soft, grainy mustard seeds.
Recipe: Aubergines *au gratin*, p. 214.

Furikake and *yukari*

Japanese cuisine makes use of a range of dried spice mixes and condiments, known as *furikake*, that can be sprinkled on cooked rice and vegetable dishes in order to add umami and a little crunch. They typically consist of roasted seaweed, sesame seeds, »chilli, »*yuzu* peel, and dried »fish flakes. *Yukari* is a particular version of *furikake* that consists of salted, dried, and granulated leaves of red *shiso* (*Perilla frutescens*), a herb belonging to the mint family. Because *shiso* has certain antifungal compounds, *yukari* is also used for conservation of vegetables and fruits, for example, pickled Japanese plums, *umeboshi*.

Flavour profile: Salty, varies depending on specific ingredients; mint aroma.
Texture: Dry, granular.
Recipe: Pickled ginger (*gari*), p. 154.

Garlic

Garlic (*Allium sativum*) is widely used to add taste and aroma to virtually any type of plant-based foods. The characteristic garlic aroma is released when the clove is chopped or crushed, causing an enzyme, alliinase, to break down the compound alliin to allicin. When heated, the enzyme is destroyed so that the odour is not formed and, at the same time, the edge is taken off garlic's sharp taste and it becomes sweeter. Garlic can turn bluish-green when it is prepared with acid because alliinase reacts with amino acids to form polypyrroles. It also has significant quantities of certain tripeptides that elicit a *koku* sensation. See also »black garlic.

Flavour profile: Sharp, strong, *koku* sensation; garlic smell from sulfur compounds.
Texture: Depends on how it is prepared.
Recipe: Tomatoes in tomatoes with mouthfeel, p. 268.

Garum

Garum is a type of »fish sauce dating back to ancient Greek and Roman times. It is produced by fermentation using salt and the fish's own intestinal enzymes to release free amino acids with umami. A number of modern variants are now being made in Europe, often using anchovies, sardines, tuna, and sprats. Similar types of sauces can also be made based on other protein-rich raw ingredients, such as insects, game, and legumes. *Garum* is an alternative to fish sauces and is used in dressings and marinades. A special mixture of *garum* (or fish sauce) with honey, so-called *meligarum*, was used in Roman cuisine, often as a dressing with vegetable dishes. A version of *meligarum* with citrus juice would then contain four basic tastes, umami, salty, sweet, and sour.

Flavour profile: Umami, *koku* sensation, salty; pungent odour.
Texture: Liquid, slightly viscous.

The Flavour Accelerators

Meligarum

½ dl (⅕ c) garum (or fish sauce)

1 dl (⅖ c) honey

1 dl (⅖ c) citrus juice, e.g., lime

1. Put all ingredients in a closed jar and shake until the honey is fully dissolved. Can be kept in the refrigerator for months.

Gastriques

Gastriques are sweet, sour, and bitter syrupy liquids made by caramelising sugar and vinegar. They can be used to add both taste and colour to a sauce and to glaze vegetables. The darker the colour of the liquid, the more bitter the taste.

Gastrique

100 g (⅔ c) sugar

1 ½ dl (⅗ c) fruit vinegar, for example apple cider, raspberry, black currant, or pomegranate

1. Caramelise the sugar in a heavy pot or skillet.
2. Warm the vinegar, pour it slowly into the caramelised sugar, and allow it to simmer until the caramel has dissolved in the vinegar. Continue to simmer the mixture until it has a syrupy consistency, but take care to prevent it from becoming too thick.

Flavour profile: Sour, sweet, bitter; vinegary smell.
Texture: Viscous, a little sticky.
Recipe: Succulent leeks, p. 240.

Ginger

Ginger (*Zingiber officinale*) is used as a spice and as a flavouring agent for all types of food, including vegetables. It contains a series of aromatic compounds, such as gingerol, zingerone, zingiberene, and shogaol. Gingerol is responsible for the sharp and burning taste sensation of ginger. Fresh ginger is sweet and tart and can be pickled to make *gari*, which is usually served with sushi and sashimi. It can also be candied or used in baking. It can be added to marinades and dressings and is a component of sodas and juices.

Flavour profile: Sharp, burning; lemon-like aroma.
Texture: Crisp to woody when raw; firm when cooked or pickled.
Recipe: Cucumbers with sesame seeds, *ponzu*, chilli, and ginger, p. 186.

Pickled ginger (*gari*)

500 g (1 lb 1 ½ oz) fresh ginger root

1 tbsp salt

375 ml (1 ⅗ c) rice vinegar

200 g (7 oz) sugar

Salted red *shiso* (*Perilla frutescens*), optional

1. Peel the ginger root and slice it very thinly using a mandoline slicer.
2. Try to keep the special contours of the ginger pieces. It is important to slice and carry out the next step very quickly to prevent the ginger from oxidising.
3. Place the slices in a bowl, pour the salt over them and massage it thoroughly into the slices.
4. Allow the salted ginger slices to rest for 1 hour in the refrigerator.
5. Drain off the accumulated liquid and place the slices on a clean cloth. Fold the cloth together and lightly squeeze out any additional liquid. Place the slices back in a bowl.
6. Warm the rice vinegar and sugar in a small pot while stirring until the sugar has dissolved and the liquid has started to boil.
7. Pour the marinade over the ginger slices.
8. Optionally, add some salted or fresh red *shiso* leaves or »*yukari* to impart a pale red colour.
9. Transfer the ginger slices and the marinade to a sterilised glass jar and refrigerate. The *gari* can keep for several months.

Goma-shio

Goma-shio is a Japanese-inspired spice mix that is easy to make using toasted, crushed sesame seeds (*goma*) and sea salt (*shio*). It can be sprinkled on vegetable dishes. See also »*za'atar*.

Flavour profile: Salty; nutty aroma due to sulfurous aromatic compounds (furfurythiol).
Texture: Crisp, oily.
Recipe: Mashed vegetables with rutabaga 'bacon', p. 208.

Gremolata

Gremolata is an Italian-inspired herb mix made with chopped parsley, lemon peel, and »garlic. It can be sprinkled on cooked vegetables.

Gremolata

1 organic lemon	1 bunch of parsley
2 cloves garlic	

1. Grate the lemon peel, chop the garlic very finely, chop the parsley, and mix them together.

Flavour profile: Sour, herbal; aromas from garlic and from volatile oils in the parsley, for example, myristicin, and limonene from the lemon peel.
Texture: Granular.
Recipe: Carrots in zucchini sauce, p. 228.

Ham, air-dried and cured

There is a large range of products made from the meat from common domesticated pigs and Iberian black foot pigs (*pata negra*). The latter forage in the forests in Spain and Portugal, where they feed on wild fruits such as acorns. These have a high content of polyunsaturated fatty acids, which is reflected in the meat of the animals, leaving it tasty and meltingly soft. During the curing process, naturally occurring moulds help to preserve the ham and produce taste and aroma substances. Only a few thin slices or small cubes of air-dried ham are needed to add a noticeable measure of umami to a dish such as a green salad. Scraps from the ham, including bits of fat, small knuckles, and rind, can be stored in an air-tight container and simmered in vegetable soups or frozen and grated for use as a spice.

Flavour profile: Umami, sweet; intense meaty smell.
Texture: Depends on the cut; the fat melts in the mouth.
Recipe: Grilled onions, p. 238.

Hemp seed crunch

Toasted hemp seeds and kernels mixed with salt make a crunchy addition that contributes an interesting mouthfeel to a whole range of vegetable dishes. Flax seeds and sesame seeds (»*goma-shio*) are also very suitable.

Hemp seed crunch

3 tbsp hemp seeds	15 g (½ oz) seaweed, for example, winged kelp or *nori*, either in pieces or granules
1 tbsp olive oil	
1 tbsp nutritional yeast flakes	1 tsp salt flakes

1. Toast the hemp seeds in the olive oil in a skillet over medium heat until they are light brown. Add the nutritional yeast flakes for the last minute or two. Pour it all out onto paper towels to absorb the excess oil.
2. Toast the seaweed on a dry skillet until it is completely dry and crush it. Mix it into the seeds together with the salt flakes. Put the crunch in a small bowl.

Flavour profile: Salty, umami.
Texture: Dry, granular, crunchy.
Recipe: Mashed vegetables with rutabaga 'bacon', p. 208.

Hoisin sauce

Hoisin sauce is a Chinese sauce typically produced from fermented soybeans, fennel, »chilli, »garlic, and possibly »vinegar and sugar. It can be used as a dip with vegetables, in marinades, to glaze vegetables, and in vegetable wok dishes.

The Flavour Accelerators

Hoisin sauce

4 small cloves of garlic	40 g (⅓ c) sesame oil
240 g (1 c) soy sauce	60 g (½ c) grated ginger root
160 g (⅔ c) red *miso* (*aka-miso*)	8 g (1 ½ tsp) chilli powder
60 g (4 tbsp) honey	20 g (4 tsp) ground pepper
20 g (4 tsp) rice vinegar	

1. Put the garlic through a press and mix all the ingredients together in a pot. Simmer to reduce them by about one-third to 3 dl (1 ¼ c).
2. Pour the sauce into a sterilised jar and seal it.

Flavour profile: Umami, sour, sweet; garlic aroma.
Texture: Viscous, sticky.

Katsuobushi

Katsuobushi is a fillet of bonito (skipjack tuna) that has been processed successively in five different ways: cooked, salted, dried, smoked, and fermented. The end result is a rock-hard fillet that resembles a piece of wood. Depending on how the fish has been caught and prepared, it has extraordinarily large amounts of inosinate that interact synergistically with umami. Ultrathin shaved flakes of *katsuobushi* are used to make »*dashi*, added to salads and dressings, and sprinkled over cold and warm vegetable dishes.

Flavour profile: Umami, salty, bitter; smoky aroma.
Texture: Dry. Thin flakes melt in the mouth.

Ketchup

Modern ketchup has its origins in a fermented Chinese fish sauce, *koe-chiap*. The fish element is long gone, and it is now made from tomato paste to which mushrooms, »anchovies, »tomatoes, »vinegar, walnuts, »pickled vegetables, and a variety of spices are added. Large amounts of sugar are also added to some commercial brands, whereby all five basic tastes are represented. It is used as a dip, often for French fries, and in dressings and sauces.

Flavour profile: Umami, salty, sour, bitter, sweet; spicy aroma.
Texture: Viscous, creamy.

Koji

Koji is a fermentation medium containing a mould, *Aspergillus oryzae*, with enzymes that can break down starches into sugar, glucose in particular, and proteins into free amino acids and peptides. The sugars can subsequently be exploited by the yeast to form a myriad of compounds, including alcohol and umami taste substances. Although it is a little tricky, it is possible to make *koji* at home using cooked soybeans, rice, barley, or another type of grain that is then seeded with spores from the mould. Fortunately, it is easier to use ready-made *shio-koji*, which contains only the required enzymes extracted from the living mould. This is also sold as a similar filtered product called *ekitai shio-koji*. Because of its high salt content, *shio-koji* keeps very well, up to one year, even without being refrigerated, so it is easy to keep it on hand. It can also be used as a taste additive, in dressings and marinades, and to make »*tsukemono*.

Flavour profile: Umami, salty; yeasty aroma.
Texture: Viscous paste, slightly granular; liquid in the case of *ekitai shio-koji*.
Recipes: *Koji*-marinated broccolini with chilli sauce, p. 178; *Koji*-marinated vegetables, p. 180.

Lees

Lees are a sediment and waste product from fermentation processes, for example, making wine or brewing »sake. The lees consist of starch, sugar, and dead yeast cells, which contain large quantities of free amino acids, especially glutamate, as well as a little residual alcohol. Thanks to their abundant umami, they can be used to flavour, marinate, pickle, and conserve vegetables and can be added to dressings. Sake lees (*sake kazu*), in particular, are sweet and not bitter. Lees from cider and wheat beer production are suitable as well, but those from other types of beer are not, due to bitter substances derived from the hops.

Flavour profile: Umami, sweet; yeasty aroma.
Texture: Granular, creamy.

Lime-soy-fish sauce dressing

Lime-soy-fish sauce dressing combines the basic tastes sour, sweet, umami, and salty with a strong »chilli taste. In addition, it has aroma substances from »garlic. The overall character of the dressing depends on the fish sauce component, so it is an advantage to experiment a little with the proportions when the dressing is to be served with a particular dish. It can be used as is on salads and vegetable dishes or form the basis for other dressings. The following recipe can easily be varied by using other citrus juices and different spices.

Lime-soy-fish sauce dressing

2 cloves of garlic	1 dl (2/5 c) sugar
A little red chilli pepper	1 dl (2/5 c) freshly squeezed lime juice
	3 tbsp Japanese soy sauce
	3 tbsp fish sauce

1. Chop the chilli pepper and garlic, add the sugar and wet ingredients, and mix them all together.
2. Refrigerate in a closed glass jar; the dressing keeps for a long time.

Flavour profile: Sour, sweet, umami, strong; aroma from the garlic.
Texture: Liquicy, granular.
Recipe: Aubergines *au gratin*, p. 214.

Marinated mushrooms

Button mushrooms or other fresh or dried mushrooms can be marinated in a little »soy sauce or »fish sauce. Before use they can be seasoned to taste with »pepper, especially marinated *sansho* pepper berries, and »citrus juice. The mushrooms can be marinated raw and added to green salads, as well as fried and baked. See also »mushroom powder and salt.

Taste and aroma: Umami, salty, earthy, possibly acidic and strong; aroma depends on the seasonings.
Texture: Soft, succulent.
Recipe: Yellow split pea hummus with thyme, p. 263.

Marmite

Marmite is the trademark for a yeast product in the form of a sticky, dark, and salty paste with a prominent umami taste. It contains significant amounts of glutamate as well as several important vitamins and minerals. Vegemite is a very similar Australian product. Both are often eaten spread on toast, mixed with boiling water to make a hot drink, or used as a taste additive in savoury dishes.

Flavour profile: Umami, yeasty; pungent aroma.
Texture: Sticky, paste-like.

The Flavour Accelerators

Mayonnaise

Mayonnaise is an emulsified sauce that, unlike hollandaise and béarnaise sauces, is eaten cold. A classic mayonnaise is made with a mixture of vegetable oil and lemon juice or »vinegar, which are emulsified with the help of an egg yolk and »mustard and seasoned with salt, »pepper, and possibly other spices. Cooking water from chickpeas (*aquafaba*) can replace the egg as a vegan emulsifier. It is made by mixing the lemon juice or vinegar with the egg yolk and then slowly adding the oil a drop at a time while stirring constantly. The mayonnaise will separate if the oil is added too quickly or if there is too little liquid. In a good mayonnaise the oil droplets are so close together that the mayonnaise becomes a little stiff and has an elastic mouthfeel. Three well-known variants, »*aïoli*, mayonnaise with »garlic, »remoulade, and tartar sauce, mayonnaise with »pickles, »capers, lemon juice, and herbs, are popular condiments with fish dishes or can be used as a dip with raw vegetables.

Flavour profile: Salty, sour, peppery; aroma from the spices.
Texture: Creamy, coats the mouth.

Mirin

Mirin is a sweet rice wine with an alcohol content of about 14 per cent. It is made from cooked rice that is inoculated with *koji*, which turns the starch in the rice into sugars, especially glucose, and the proteins into free amino acids, especially glutamate, with umami taste. Yeast is then added to ferment the sugars into alcohol. *Mirin* is not drunk but is used to add sweetness and umami to food. It is an ingredient in »*ponzu* and »*sanbaizu*. It can be used for steaming, glazing, and caramelising vegetables and added to dressings and sauces.

Flavour profile: Sweet, umami; alcoholic and slightly fruity aroma.
Texture: Liquid.
Recipes: *Ponzu* and *sanbaizu* p. 162. Seaweed 'liquorice', p. 165.

Miso

Miso is a paste made by fermenting soybeans and/or different grains with the help of »*koji*. It typically has a 14 per cent protein content, as well as large quantities of free amino acids, especially glutamate. The salt content varies from 5 to 15 per cent. It becomes darker and stronger as a function of a longer fermentation and storage period. *Shiro-miso* is sweet, white *miso*; *aka-miso* is red *miso*; *mugi-miso* is made with barley and slightly sweet; *miso-zuke* are vegetables pickled in *miso*; *miso* soup is made by adding *miso* to »*dashi*. *Miso* is one of the most important ingredients in vegan Japanese temple cuisine, *shōjin ryōri*. It can be used in soups, as a dip, and in dressings, sauces, and marinades.

Flavour profile: Umami, salty, sweet in the case of *shiro-miso*; yeasty aroma.
Texture: Creamy or granulated, depending on the type.
Recipes: *Miso* soup with seaweed, p. 188. Deep-fried aubergines, p. 210. Aubergines *au gratin*, p. 214. Broccoli with *miso*-mayonnaise, p. 222.

Miso mayonnaise

2 egg yolks	2 tbsp apple cider vinegar
1 tsp *shiro-miso*	Salt
1 tsp mustard	2 dl (⅘ c) neutral-tasting oil

1. All ingredients should be at room temperature.
2. In a deep bowl beat together the egg yolks, *miso*, mustard, vinegar, and salt.
3. Beat in the oil drop by drop at first, then a little at a time until the desired consistency is reached.

Mushroom powder

Mushrooms, for example, *shiitake*, funnel chanterelles, porcini, and other dark, edible fungi can be dried in a dehydrator or an oven at a low temperature, during which time guanylate, which is the basis for umami synergy, is formed. This concentrates the umami taste, especially in *shiitake*. The mushrooms are subsequently ground into a powder or fine granulate and mixed with salt flakes. The powder is a fine flavour additive for vegetable dishes and sauces, dressings, and soups.

Flavour profile: Umami, salty, sweet; earthy aroma.
Texture: Dry, granular
Recipe: Yellow split pea hummus with thyme, p. 263.

Mustard

Mustard is prepared from the seeds of mustard plants. White or yellow seeds are from *Sinapis alba*, brown from *Brassica juncea*, and black from *Brassica nigra*. The seeds are ground coarsely or finely with water, vinegar, salt, and possibly sugar and herbs such as tarragon and horseradish. Oil in the mustard seeds contain certain proteins that help to emulsify the mixture, turning it into a soft paste. Crushing the seeds releases an enzyme that produces isothiocyanate, which is responsible for the sharp, burning, and irritating taste. The more acidic the mustard, the longer this sensation lingers in the mouth. Dark seeds have a stronger taste than the light-coloured ones. Mustard is used as a condiment on its own and in dressings, sauces, and marinades.

Flavour profile: Sharp, burning, irritating, possibly sweet; vinegary aroma.
Texture: Creamy, can be granular.
Recipes: *Aïoli*, p. 147. Fruit vinegar marinade, p. 153. Celeriac salad, p. 182.

Nutritional yeast

Nutritional yeast is made from dried and inactivated baker's yeast (*Saccharomyces cerevisiae*) that is cultivated on a substrate of molasses from sugar beets or sugar cane. The yeast cells and their enzymes are deactivated by heating, after which the yeast is washed and dried. The finished product is yellowish, in the form of either a powder or small flakes. Nutritional yeast is different from yeast extract, which is darker, has a much stronger taste, and a higher glutamate content. Because nutritional yeast is a good source of vitamin B and all the important amino acids it is often recommended as a nutritional supplement. Yeast flakes can be sprinkled on vegetable dishes and popcorn. When dispersed in cold water or sake, yeast flakes make a good, creamy dressing with umami tastes, possibly with the addition of a little vinegar and some spices or seaweed granulates.

Flavour profile: Umami, nutty taste similar to that of Parmesan cheese and *shiro-miso*; yeasty aroma.
Texture: Dry, flaky, creamy when suspended in water.
Recipes: Green asparagus, p. 220. Celeriac salad, p. 182.

If living yeast cells are broken open, the enzymes in the cells can break down their protein content by hydrolysis, thereby releasing large quantities of glutamate. This action is the basis for the industrial production of yeast extract, which is heavily used to add umami to commercial products. »Marmite and Vegemite, which are completely vegan spreads, are well-known examples.

The following recipe for a dressing using nutritional yeast comes from an old hippie café in Florida, which was very popular with vegetarians and vegans. The dressing can be used on green salads or as a dip, for example for French fries or raw vegetable snacks.

Dressing with nutritional yeast

10 g (2 tsp) nutritional yeast flakes	¼ dl (1 ⅔ tbsp) apple cider vinegar
½ dl (⅕ c) tamari soy sauce	½ tsp thyme
½ dl (⅕ c) grapeseed oil	½ tsp dried basil
¾ dl (⅓ c) lemon juice	Black pepper

1. Stir all ingredients together. Seaweed flakes can be substituted for the herbs.
2. To make a lighter dressing without oil, mix the nutritional yeast flakes with rice vinegar and seaweed flakes.

Oyster sauce

Genuine oyster sauce is prepared by reducing the water in which oysters have been cooked. The result is a caramelised, brown liquid to which salt is added. Some commercial products are made with an oyster extract that is thickened with maize starch, coloured with caramel, and seasoned with salt and MSG. It is used in stir-fries and in dressings for vegetables and beans.

Flavour profile: Umami, salty; musky, fishy odour.
Texture: Viscous.
Recipe: Sautéed spinach, mung beans, and wood ear mushrooms, p. 192.

Panko

Panko are flaky, very dry Japanese breadcrumbs that are particularly well-suited for breading. They are made from a special kind of bread that is baked by passing an electric current through wheat dough that has been proofed several times. The baked bread is very airy and has no crust. When it is completely dry it is shaved into fine slivers and made into granules. *Panko* has a distinctive porous structure that limits the amount of oil it will absorb while being fried ensuring that the breading becomes very crisp, as in Japanese tempura.

Flavour profile: Sweet; mild bread aromas.
Texture: Dry, crispy, crunchy.
Recipes: Green asparagus, p. 220. Polenta fritters with blue cheese dip, p. 256. Deep-fried aubergines, p. 210. Anchovy sauce (*bagna càuda*), p. 148.

Parmesan cheese

Parmesan cheese (Parmigiano-Reggiano) is a hard, granulated dry cheese, produced from a mixture of whole and skimmed unpasteurised cow's milk. It has both low salt and fat contents. The cheese is aged for at least two years, during which time large quantities of glutamate develop. Parmesan cheese is one of the processed food products that contains the most glutamate, dwarfed only by some »fish sauces. Its strong umami can help to suppress the bitter taste of some vegetables, while the glutamate brings out sweetness. The cheese rinds and small leftover bits can be stored in a container in the refrigerator to be added to soups and sauces while they are cooking and then discarded.

Flavour profile: Sharp, umami, salty, bitter, and with *koku* sensation; pungent odour caused by butyric acid can be off-putting.
Texture: Firm, dry, granular.
Recipes: 'Greens to go' casserole, p. 202. Kale pancakes, p. 196; Onion soup—the real thing! p. 190. Aubergine 'fettucine,' p. 216. Grilled onions, p. 238.

Pepper

There are many species of genuine pepper from the *Piper* genus. The most widely used types are made from the berries of *Piper nigrum* plants. Black pepper is made from unripe whole berries that are cooked and dried to make small peppercorns. Green or Madagascar peppercorns are made from unripe berries that are then freeze-dried or treated with sulfur dioxide to preserve their colour. White pepper is made from fully ripe berries that are fermented by soaking in water, after which the outer layer is stripped away. *Piper longum* is a close relative of *Piper nigrum* but is hotter. It consists of tiny fruits, similar in size to poppy seeds that are attached to the surface of a flower spike. A number of spices that are also thought of as peppers do not belong to the *Piper* genus. Examples are pink peppercorns (*Schinus molle*), all spice (*Pimenta dioica*), sansho (*Zanthoxylum piperitum*), grains of paradise (*Aframomum melegueta*), and Sichuan pepper (*Zanthoxylum simulans*). All of these peppers, whether or not they are the 'true' ones, lead to varying degrees of hot and burning taste impressions and are differentiated by their distinct aroma profiles. They are used universally to add heat to vegetables dishes, dressings, marinades, and sauces. Black pepper owes its special aroma to the presence of more than a hundred volatile compounds. The one that is responsible for its characteristic burning taste impression is piperine, which is found in larger quantities in black than in white pepper. In addition, there is another substance, chavicine, which is hotter than piperine and is its geometric isomer. Both of these compounds are irritants in that they have a chemesthetic effect on skin and mucous membranes that causes pain and can potentially damage cells and tissues. So, what we think of as a 'pepper taste,' is not a true taste, but instead a burning and irritating mouthfeel.

Flavour profile: Strong, burning; various aromas.
Texture: Crushed and granular or powdery.

Kampot pepper, a very aromatic variant of black pepper, is featured extensively in Cambodian cuisine. It is an ingredient in a pepper sauce used to add flavour to dishes such as Green fricassee, p. 204, and *Koji*-marinated vegetables, p. 180. Although it can come across as a bit rustic, it makes an ideal dip for individual servings of vegetables.

Kampot pepper sauce

¼ dl (1 ⅔ tbsp) Kampot peppercorns	Rind from 1 organic lime
¾ dl (5 tbsp) nutritional yeast flakes	½ dl (3 ⅓ tbsp) freshly squeezed lime juice
	2 g (½ tsp) sea salt

1. Crush the peppercorns in a mortar or a spice mill.
2. Toast the nutritional yeast flakes until they are light brown.
3. Grate the rind of the lime using a microplane. Squeeze the lime juice.
4. Mix everything together with the sea salt. Keep refrigerated for later use.

Pink peppercorn sauce

25 g (1 ⅔ tbsp) dark brown sugar	¼ dl (1 ⅔ tbsp) neutral-tasting oil
25 g (1 ⅔ tbsp) Dijon mustard	1 tsp fish sauce
1 g (¼ tsp) crushed pink peppercorns	½ tsp rice vinegar
	2 tsp soy sauce

1. Mix all the ingredients together.
2. Keep refrigerated for later use, for example, with Dry-cured carrots, p. 228.

Pesto

Pesto is a purée, paste, or sauce that is partly an oil emulsion. Classical Italian pesto is made by coarsely pulsing together fresh basil leaves, »garlic, and pine nuts in a food processor and then stirring in olive oil and, possibly, »Parmesan cheese. The basil leaves should be just big enough to require a bit of chewing, but small enough to have a soft mouthfeel and to keep the pesto from separating. It can be spread on bread and used as a sauce or a dip. Parsley and seaweeds can be substituted for all or some of the basil. The following recipe for a seaweed pesto is from Anita Dietz.

Seaweed pesto

20 g (¾ oz) dried seaweeds (for example, sugar kelp, serrated wrack, winged kelp, tangle)	20 g (2 ¼ tbsp) capers
	A little fresh parsley or spinach
1 avocado	2 tbsp olive oil
1 red onion, peeled and cut into large pieces	Salt and pepper
1 clove of garlic, peeled	Ca. 75 g (6 tbsp) pumpkin seeds

1. Simmer the seaweed in a little water for 10 minutes. Drain the seaweed.
2. Peel the avocado. Cut it up into a few pieces.
3. Blend everything together.
4. Before serving, mix in a few whole pumpkin seeds to add texture.

Flavour profile: Slightly bitter, umami, sour; garlic smell, other aromas according to the ingredients.
Texture: Creamy, oily, granular.

Horseradish-parsley pesto with capers

30 g (⅘ c) parsley	40 g (3 ¼ tbsp) capers
1 clove of garlic	1 dl (⅖ c) oil
3–5 g (½–1 tsp) grated horseradish	Pinch of salt

1. Chop the parsley coarsely, cut the garlic up into small pieces, grate the horseradish, and blend all together with capers and oil.
2. Season with salt to taste.

Flavour profile: Salty, strong, acidic; herbal parsley aroma.
Texture: Soft, granular.
Recipe: 'Greens to go' casserole, p. 202.

The Flavour Accelerators

Pickles

Pickles are made from vegetables using salt or vinegar, and possibly some sugar and spices such as chilli and dill. They can be served as a condiment or a small dish on their own or used to add taste and texture to other plant-based dishes. Pickles are often made with various types of cucumber, and sometimes lactic acid fermentation is also involved. Cornichons, also called gherkins, are pickled baby cucumbers that are spiced with tarragon. Chopped pickles are added to relishes. »*Tsukemono* are Japanese pickles made from a wide assortment of vegetables and even a few fruits and flowers.

Flavour profile: Sour, salty, possibly sweet and spicy; vinegary aroma.
Texture: Crisp, juicy.

Ponzu and *sanbaizu*

Sour marinades tempered with sweetness and umami pair well with vegetables. Japanese cuisine is world-famous for marinades or condiments of this kind, all made with flavourful rice vinegar (*su*). The best-known are *ponzu* and *sanbaizu*, which are incorporated into dressings, marinades, and sauces. The recipes for making them that follow below are based on »*dashi*, the Japanese soup stock made with an extract of *konbu* (see »seaweeds) and a fish product called »*katsuobushi*. The liquids keep for a long time stored in a bottle.

Ponzu and *sanbaizu*

Makes ca. 215 ml (9/10 c)	1 tbsp rice vinegar
4 tbsp Japanese soy sauce	3 tbsp *dashi*
6 tbsp *yuzu* juice or lemon juice	2 tsp *mirin*

1. Mix all the ingredients together to make *ponzu*.
2. Leave out the *yuzu* or lemon juice to make *sanbaizu*.

Flavour profile: Umami, sour, salty: citrus aromas from *yuzu*.
Texture: Liquid.
Recipes: Cucumber with sesame, *ponzu*, chilli, and ginger, p. 186. Kidney bean and *tsukemono* salad, p. 180.

Potato cooking water

Water left over from boiling potatoes is a sadly overlooked source of umami, especially when it is from potatoes that are unpeeled or so old that they are about to sprout. It has a fair amount of glutamate and can be used to steam or sauté other vegetables or added to sauces, stews, and soups. The taste of the water depends on the variety of potato that was cooked in it.

Flavour profile: Umami, sweet; bland, earthy aroma.
Texture: Liquid, slightly starchy.

Remoulade

Remoulade is a »mayonnaise-based sauce with a variety of chopped sweet and sour »pickles, such as »*tsukemono*, as well as herbs, possibly other vegetables, and spices. Its texture combines the creaminess of mayonnaise with crispness and crunch from the pickles. It is typically eaten as a cold condiment, often as a dip, or with fish dishes, French fries, and roast beef.

Flavour profile: Sour, umami, possibly sweet depending on the pickle content; vinegary smell, aromas vary according to ingredients.
Texture: Creamy, granular from the chopped pickles.

Roasted nuts and seeds

Roasted seeds from pumpkins, sunflowers, hemp, and sesame seeds, as well as nuts such as cashews or hazelnuts have tasty oils and become crisp when toasted on a pan or roasted in the oven. Adding some soy sauce or »*ponzu* toward the end of the process introduces a wealth of umami. The seeds and nuts contribute interesting textures and tastes to salads and vegetable dishes.

Flavour profile: Possibly umami, depending on the ingredients; aromatic oils.

Texture: Crisp, crunchy.

Recipes: 'Greens to go' casserole, p. 202. Tomatoes in tomatoes, with mouthfeel, p. 268. Sweet potatoes, p. 250.

Seeds with soy sauce

2 dl (4/5 c) pumpkin or sunflower seeds	A little oil
	½ dl (⅕ c) soy sauce

1. Put the seeds in a cold skillet with a little oil. Warm up to medium heat and toast the seeds, stirring frequently, until the seeds start to puff up.
2. Add the soy sauce and toss the seeds in it. Remove them and place on paper towels to absorb excess oil. Sprinkle on a dish just before serving it.

Tamarind cashew nuts

100 g (3 ½ oz) cashew nuts	5 g (1 tsp) sea salt
1 dl (⅖ c) water	1 tbsp tamarind sauce

1. Bring the cashew nuts, salt, and water to a boil. Turn off the heat, let the nuts stand in the water for 20 minutes, and then drain.
2. Coat the nuts with the tamarind sauce and spread them out on parchment paper in a baking pan.
3. Bake the nuts for ca. 20 minutes at 140°C (285°F), tossing them a few times.
4. The cashews will keep for a long time in a tightly closed container. A sample recipe is: Celeriac salad, p. 182.

Salted, smoked peanuts

100 g (3 ½ oz) salted peanuts	Smoking chips

1. Put the smoking chips in a smoker. Spread the peanuts out on a disposable roasting pan and place in the smoker.
2. Smoke at low heat for 10 minutes. Then toast them on a dry skillet until they are golden.

Romesco sauce

Romesco sauce is a traditional Catalan tomato-based sauce that originated in the Tarragona region of Spain. It was first made several centuries ago by the fishermen as a simple condiment with »garlic, dry peppers, oil, bread, and wine to be eaten with their local catch. Over time, »tomatoes were introduced and became a distinctive feature of the modern version. There are many variations of this popular condiment, which is widely used on meat and fish dishes, with grilled vegetables, and as a dip.

Romesco sauce

1 red bell pepper (or 5 *choriceros*)	½ dl (3 ½ tbsp) red wine vinegar
2 medium size tomatoes	1 ⅕ dl (½ c) olive oil
2 large cloves of garlic	1 slice country bread without crust, about 50 g (1 ¾ oz), torn to small pieces
20 g (¾ oz) blanched almonds	
10 g (⅜ oz) hazelnuts	Salt
	Freshly ground pepper

1. If using *choriceros* (dehydrated red peppers), cook them for 3 minutes, then remove and discard the skin and seeds.
2. Turn on the oven to 185°C (365°F).
3. Cover a rimmed baking sheet with parchment paper and place the fresh bell pepper, tomatoes, almonds, hazelnuts, and garlic on it. These ingredients need different cooking times.
4. Place the baking sheet in the oven. Remove the almonds and hazelnuts when they are light brown. Rub the hazelnuts between the hands to remove the brown outer shell.
5. Remove the tomatoes and garlic after about 10–15 minutes and the bell pepper 5 minutes later.
6. Put the pepper in a brown bag to cool. Then peel off the skin and remove the seeds and discard them.
7. Using a hand blender combine these ingredients plus the wine vinegar. To make a coarser *romesco*, take out about ¼ dl (1 ½–1 ¾ tbsp) of the liquid and put it aside.
8. Blend in the oil and the bread pieces a little at a time until the sauce is thick and creamy or the desired consistency. This may not require all the bread. Stir into the liquid that was set aside.

The Flavour Accelerators

9. Season to taste with salt and pepper, and possibly a little sugar and wine vinegar.

Flavour profile: Sweet, strong, garlic, acidy, umami, *koku* sensation; smoky aromas.
Texture: Liquid, granular texture from the nuts.
Recipe: The use of *romesco* sauce to accompany traditional Catalan preparations of *calçots* is described on p. 20. *Calçots* are onions that are like a cross between a leek and a spring onion. When they are charred on a grill, they take on a special sweetness that is quite different from that of cooked leeks.

Rouille

Rouille is a sauce that can be thickened with breadcrumbs or the crusts from stale bread. It is made by mixing together olive oil with »chilli or cayenne pepper, crushed »garlic, and saffron and then stirring in breadcrumbs to achieve the desired consistency. It can also be made as a variation on » *aïoli* to which breadcrumbs have been added. It is used in fish soups and spread on bread.

Flavour profile: Spicy, *koku* sensation; garlic taste and aroma.
Texture: Creamy, slightly granular.

Sake

Sake is rice wine that is produced by fermenting cooked white rice using »*koji*, which has enzymes that can break starch down into sugar and proteins and release free amino acids, giving rise to sweetness and umami. A yeast culture then converts the sugar to alcohol. Sake can add umami when steaming vegetables and in sauces and dressings. The lees from the brewing process, *sake kazu*, (see »lees) which has a considerable content of dead yeast cells and amino acids, is a very potent source of umami and makes an excellent pickling bed for vegetables. Cooking sake is a lower quality product that is intended solely for cooking and not for drinking. Sake is used more widely in Japanese cuisine than either beer or wine in Western kitchens.

Flavour profile: Sweet, umami; wine-like aromas.
Texture: Liquid.
Recipe: Basic *tsukemono* marinade, p. 167.

Salsa verde

Salsa verde is a spicy, green sauce that is featured in Mexican cuisine. Traditionally it is made by blending together tomatillos (*Physalis* spp.) that are either raw, roasted, or cooked and »chilli peppers. Oil may then be added. Some versions are seasoned with parsley, onions, »garlic, »anchovies, »capers, and possibly »mustard. Anchovies and capers are a source of umami and garlic provides a *koku* sensation. *Salsa verde* is used as a dip with vegetables and in sauces, dressings, and soups.

Flavour profile: Sweet, strong, possibly umami and *koku* sensations; herbal aroma.
Texture: Sauce-like consistency; thickness depends on how it was prepared.

Sansho

Sansho is a Japanese variety of pepper, *Zanthoxylum piperitum*, that is not a true pepper. Its taste and aroma are similar to Chinese Sichuan pepper. It is used in Japanese cuisine primarily to offset the aroma of fats but can also serve as a condiment for plant-based dishes. *Sansho* is available whole as dry seed shells for grinding and as a powder. The seed shells can be simmered in »soy sauce. The hot taste sensation of *sansho* is more subtle and aromatic than that of black »pepper and »chilli. It is a component of the spice blend, »*shichimi*. It is commonly used in dressings and sauces and as a flavouring for marinated mushrooms.

Flavour profile: Hot, burning; aromas similar to those of mint, basil, and lemon, notes of liquorice.
Texture: Granular or powdery, depending on how it is ground.

'Seaweed liquorice'

Brown seaweeds such as sugar kelp (*Saccharina latissima*), tangle/oarweed (*Laminaria digitata*), and preferably *konbu* (*Saccharina japonica*) can be prepared to take on a texture similar to that of soft liquorice and it has a powerful umami taste. The 'liquorice', called *konbu tsukedani* (if prepared with *konbu*), can be eaten as a snack or added to both cold and warm vegetable dishes. The recipe below is very adaptable, so it does not specify quantities.

'Seaweed liquorice'

Dried brown seaweeds, for example, sugar kelp (*Saccharina latissima*), tangle (*Laminaria digitata*), konbu (*Saccharina japonica*)

Soy sauce

Mirin or sugar

Shiitake powder

Salt

Rice or soy flour

Possibly a little sal ammoniac

1. Soak the dried seaweeds in a little cold water.
2. Simmer the seaweeds for a little less than an hour in a small amount of water mixed with soy sauce, *shiitake* powder, and *mirin* (or sugar). As the liquid should not be allowed to boil away, add water as necessary.
3. Remove the seaweeds from the liquid and place on paper towels to dry for about two hours. Blot them to absorb the slimiest surface water.
4. Cut the seaweeds into 5–10 cm (2–4 in) lengths.
5. Make a mixture of salt and rice or soy flour (possibly with a little sal ammoniac if the 'liquorice' is to taste salty) and place the mixture in a large plastic bag.
6. Put the seaweeds in the plastic bag and shake until the surfaces are dusted with a fine layer of the mixture so that they do not stick together Skip this step and the next one if the 'liquorice' is to have a sticky mouthfeel.
7. Cut into appropriately sized pieces.

Flavour profile: Umami, salty; marine aroma.
Texture: Chewy, similar to soft liquorice.

Seaweed powder and granulate

Seaweeds in the form of powders and granules can be used as salt substitutes and flavour enhancers. These will contribute different tastes and degrees of saltiness depending on whether they are composed of brown, red, or green varieties of seaweeds either on their own or mixed together. The seaweeds are usually dried at about 40°C (105°F) in order to retain as much of their vitamin content and colour as possible. If stored in an air-tight container or a jar with a tight lid, the powders and granulates will keep for years. They can be sprinkled on vegetable dishes and added to bread doughs, dressings, sauces, and soups.

Flavour profile: Umami, salt, bitter; depending on the species may have aroma notes of sulfur, iodine, or bromine.
Texture: Dry, granular.
Recipe: Mashed vegetables with rutabaga 'bacon', p. 208.

Shellfish powder

Shellfish powder can be made with dried mussels, shrimp, and littoral crab. Water from cooking blue mussels (*Mytilus edulis*) can be reduced to make mussel powder. Whole shrimp heads can be smoked, dried and ground in a spice mill to make shrimp powder. When mixed with potato cooking water or liquid squeezed from sun-ripened tomatoes, the shrimp powder makes a fantastic »*dashi* with a slightly smoky taste and good umami synergy from the inosinate content. Crab shells are cleaned and sterilised, then dried and ground to a powder. Shellfish powders can be dissolved in water to make a light broth and they can also be stirred into sauces and dressings.

Flavour profile: Umami, salty; smell of the sea, possibly smoky.
Texture: Powdery.

The Flavour Accelerators

Shichimi

Shichimi is a special Japanese spice blend that typically includes »*sancho*, red chilli, dried »ginger, sesame seeds, *ao-nori* (green seaweed flakes), »*yuzu* peel, and hemp seeds. There are different local variations, but they are always made from seven ingredients. *Shichimi* can be used as a general spice, much like pepper, and in sauces, marinades, and dressings.

Flavour profile: Umami, sour, strong, sharp; aromatic oils.
Texture: Dry, possibly crunchy.

Soy sauce

Soy sauce is made by fermenting soybeans, wheat, and possibly other cereals such as rice and barley. The end products reflect the flavour characteristics of the combination of basic ingredients and have a high salt content of 14–18 per cent. Sauce made from wheat and rice is sweeter. The secret behind the brewing process is the fermentation medium »*koji*. A yeast culture and lactic acid bacteria, which thrive in the very salty environment, either arise spontaneously or are added. The activity of the enzymes and microorganisms breaks down fats and proteins to large quantities of umami-rich glutamates. Because the brewing process and subsequent aeging periods are very lengthy, Maillard compounds with a rich flavour profile are formed. Japanese soy sauces are preferred to the Chinese ones for seasoning vegetables and in dips and marinades because they are lighter, less viscous, less sweet, and have a more delicate taste.

Flavour profile: Umami, salty; tastes and aromas from Maillard compounds.
Texture: Liquid.
Recipe: »*Ponzu*, p. 162.

Tahini

Tahini is a paste made from ground sesame seeds. Since sesame seeds are very rich in oils and also have a high antioxidant content, tahini can keep for a long time without becoming rancid. Because the seeds have a little free glutamate, the paste can contribute some umami to dressings and vegetables, especially if mixed with »Worcestershire sauce and »*miso*. It is also used to make many Middle Eastern and Asian sweets.

Flavour profile: Umami; nutty aroma (from furfurylthiol) similar to that of roasted coffee.
Texture: Oily, creamy.
Recipes: Baked root vegetables, p. 230. Deep-fried aubergines, p. 210.

Tamarind sauce

The pod-like fruits of a leguminous tropical tree, *Tamarindus indica*, are made into a sour, sweet sticky paste or sold as a solid stick or cake. It is an ingredient in »Worcestershire sauce and is often used in curries and chutneys or diluted to flavour sauces and dressings for vegetable dishes.

Flavour profile: Sour, sweet; aromatic roasted taste.
Texture: Sticky, paste-like.
Recipe: Tamarind cashew nuts, p. 163.

Tomatoes, sun-dried and in paste

The most intense tomato taste is found in dried, sun-ripened tomatoes, tomato purée, and in tomato paste. These ingredients are an almost universal means of imparting umami to vegetable dishes of all kinds. The paste is also used in dressings, sauces, and marinades.

Flavour profile: Umami, sweet; complex tomato aromas.
Texture: Firm and a little chewy or paste-like.
Recipes: Aubergine 'fettucine,' p. 216; Simmered tomato sauce with herbs and vegetables, p. 167.

Simmering or baking sun-ripened tomatoes slowly over several hours greatly concentrates their umami potential. If chopped up vegetables and herbs are mixed in with the tomatoes the resulting texture is a little lumpy and the sauce can almost be mistaken for a Bolognese sauce with minced meat.

Simmered tomato sauce with herbs and vegetables

1 kg (4 1/5 lb) greens, such as onions, celery, parsley root, carrots, leeks

2 kg (8 2/5 lb) very sun-ripened blanched tomatoes (canned tomatoes can also be used)

1 dl (2/5 c) olive oil

4 cloves of garlic

1 chilli

A handful of fresh thyme sprigs

4–5 bay leaves

50 g (4 tbsp) tomato paste

Salt

Freshly ground pepper

1. Coarsely chop up the vegetables, garlic, and tomatoes in a food processor or meat grinder. Warm the olive oil in a heavy pot with a tight-fitting lid and stir in the vegetable mix.
2. Strip the thyme leaves from the stalks, chop them, and add them to the pot together with the tomato paste and the bay leaves.
3. Cover the pot with parchment paper and the lid. Simmer on the stove at very low heat or in an oven at 140° (285°F) for 3–4 hours—the longer, the better.
4. Season with salt and pepper and store in canning jars in the refrigerator or freezer.
5. When the sauce is to be used it should be warmed to the boiling point. Then butter, about 50 g (3 oz) butter per 500 g (2 c) of the resulting tomato sauce, should be whisked in—more butter gives a fuller taste. It can also be seasoned with 5–10 leaves of freshly chopped sage to taste.
6. Serve with pasta and grated Parmesan cheese.

Taste and aroma: Umami, sweet, *koku* sensation; aromas depend on the overall combination of the ingredients.
Texture: Creamy, granular.
Recipe: Green 'lasagna,' p. 218.

Tsukemono

Tsukemono are preserved and pickled vegetables prepared according to centuries-old Japanese techniques. The processes involve drying, marinating, salting, pickling, and fermenting, either alone or in combination, and variously make use of salt, vinegar, sugar, alcohol, *dashi*, and a range of spices. *Tsukemono* are eaten as a snack or condiment or added to salads and relishes. The basic marinade described below can be seasoned according to taste.

Flavour profile: Sour, salt, sweet; possibly yeasty aroma.
Texture: Crisp, crunchy.
Recipes: *Koji*-marinated broccolini with chilli sauce, p. 178. Kohlrabi *tsukemono*, p. 173.

Basic *tsukemono* marinade

5 dl (2 1/8 c) water

17 1/2 g (3 1/2 tsp) sea salt

10 g (2 tsp) sugar

1 tsp lemon juice

5 g *konbu*

1. Mix together water, salt, sugar, and lemon juice and stir thoroughly or heat until the sugar and salt have dissolved completely. Immerse the *konbu* in the liquid.
2. Cool the marinade and then season it according to how it is to be used, for example, with peppercorns, sake, wine vinegar, *ponzu*, citrus juice, or herbs.
3. Store in the refrigerator until it is to be used.
4. Pour the marinade over the vegetables that are to be preserved, covering them completely.

Verjus

Verjus literally means 'green juice' and is made by cooking unripe grapes and reducing the liquid to a syrupy juice, which is somewhat sharp, sweet, and sour, and has a herbal aroma. It can be used as a substitute for »vinegar and lemon juice in plant-based dishes. »Balsamic vinegar can be viewed as *verjus* that has been fermented and aged for a long time. *Verjus* is used to deglaze and in sauces and dressings.

Flavour profile: Sour, sweet, sharp; herbal aroma.
Texture: Viscous.

Vinaigrette

Vinaigrette is a cold, very fluid sauce or dressing that is a mixture of oil and vinegar whisked together with salt, »pepper, and possibly a variety of spices and other ingredients. These could be, for example, »mustard, lemon juice, or »tomato purée. In its simplest form, the oil and vinegar do not form an emulsion and the two liquids partially separate. When the vinaigrette is shaken, it thickens a little because the oil forms smaller droplets. It, therefore, has to be shaken vigorously before use each time.

Flavour profile: Salty, sour, possibly hot and spicy; aromas from vinegar and spices.
Texture: Liquid, sometimes separates into oil and vinegar components.

Vincotto

Vincotto is an Italian specialty that literally means 'cooked wine.' It is a thick paste made by the slow cooking of freshly crushed grapes to reduce the must to about one-fifth of its original volume and caramelise the sugars. It tastes much like balsamic vinegar, but is much less sour. It is an excellent addition to dressings and can be used on cooked vegetables.

Flavour profile: Sweet; fruity aromas with caramel notes.
Texture: Viscous, paste-like.

Vinegar

Vinegar is made from wine or a fermented fruit juice, such as apple cider, by acetic fermentation to turn the alcohol into acetic acid along with a cascade of aromatic compounds that reflect its origins. Vinegar is the most important local source of sour tastes in food cultures where »citrus fruits are not grown. »Balsamic vinegar is aged in wood barrels to develop a richer flavour profile. Japanese rice vinegar (*su*) is made from rice wine and has a softer and more delicate taste than most vinegars based on fruits. Vinegar is used in dressings, marinades, pickles, and preserves.

Flavour profile: Sour; a variety of aroma substances depending on ingredients and how it is fermented.
Texture: Liquid.

Worcestershire sauce

Worcestershire sauce was originally a type of fermented anchovy sauce related to the classical Roman fish sauce »*garum*. It is made from »anchovies, wine vinegar, molasses, salt, sugar, »tamarind, »onions, »garlic, and a variety of optional flavour additives, such as »soy sauce, cloves, lemons, »pickles, and »peppers. It can be used in dressings, stews, and other dishes to add a burst of spiciness.

Flavour profile: Umami, salty, sweet, sour; aromatics from spice additives.
Texture: Liquid to slightly viscous.
Recipes: Radicchio, p. 184. Citrus dressing, p. 151.

Yuzu

Yuzu is a small Japanese citrus fruit, *Citrus junus*, with a more complex taste and aroma than lemons and limes. The most intense flavours are in the rind, whereas the juice is somewhat bitter. The juice and rind are used in »*ponzu* and »*shichimi*. The juice can be added to dressings, marinades, and sauces for vegetable dishes.

Flavour profile: Sour; citrus-like, complex aroma.
Texture: Depends on how it is prepared.

Za'atar

Za'atar is a mixture of culinary herbs, such as oregano, thyme, and savory, with seeds, sumac, and salt. It is very popular in the cuisines of the Mediterranean region of the Middle East, where it is used in dressings, on vegetables, and on freshly baked flat breads.

Za'atar

- ½ dl (⅕ c) sesame seeds
- 2 tsp coriander seeds
- ½ dl (⅕ c) dried sumac
- 1 tsp dried oregano
- ½ tsp sea salt
- ½ tsp thyme
- A little chilli powder, according to taste

1. Roast the sesame and coriander seeds on a dry skillet until they are light brown.
2. Grind the seeds and sumac in a spice mill or crush them with a mortar and pestle.
3. Mix with the rest of the ingredients, and season with salt to balance out the acidity.
4. Store in a tightly sealed jar.

Flavour profile: Salty, sour; nutty and herbal aromas.
Texture: Dry, granular.
Recipes: Couscous, p. 252. Mashed vegetables with rutabaga 'bacon.' p. 208.

RECIPES

This chapter has a large selection of recipes that are examples of how to use the principles already described to enhance the taste and texture of plant-forward dishes. Many of the recipes make use of *The Flavour Accelerators* described in the previous chapter.

The recipes range widely in terms of raw ingredients, preparation methods, and tastes. To a large extent they reflect the authors' own interests and their wish to show their readers how easy it is to prepare delicious green cuisine. In particular, they highlight ways to add or enhance umami, that all-important taste that is so often lacking in vegetables. It took about two years to finalise the recipes in order to be able to move through the different seasons one or two times. This long time span also made it possible to be very adventurous in trying out new and 'crazy' ideas and to handle certain raw ingredients in a different way the second time around.

As stated at the outset, this book is not a vegetarian or vegan cookbook, but rather a guide for people who wish to follow a diet with a greater emphasis on plant-based food. Consequently, some of the recipes incorporate small amounts of meat, fish, dairy products, and eggs, as well as those Flavour Accelerators that are based on fermented fish and shellfish. But in all cases the focal point of the dishes has been turned around, so to speak, in that the plant-based ingredients occupy centre stage, replacing the meat that is often featured most prominently in some food cultures.

Part of the idea behind the book is to present versatile recipes that can be served as appetisers, entrées, or side dishes. Many of them can be combined to make up a two or three course menu. In these cases, it may be necessary to scale them up or down in order not to prepare too much or too little food. And some of the dishes can easily be kept refrigerated and eaten a day or two later, possibly with the addition of new raw ingredients and taste substances. This helps to minimise food waste and save preparation time.

When they are cooking, some people prefer to follow recipes exactly using only the specified ingredients. They can simply use our recipes in that way. Sometimes, however, an ingredient may not be on hand, may not be in season, or may be hard to find in a particular area. When that happens, we encourage readers to experiment by using the principles in earlier chapters to achieve the desired taste and texture. In many cases, there are options for replacing a raw ingredient with something that will produce a similar result and we have suggested some of them. It is, to a certain extent, also a matter of being open-minded and a bit adventurous. Also, because metric measurements do not always convert neatly to imperial ones, we have chosen to use approximations that are sensible and sufficiently precise in a culinary context.

The recipe section starts off with some simple pickles that provide an easy way to introduce new flavours and add more vegetables to just about any meal. The others are grouped by ingredients or type of dish, including a few that are quite challenging. Bon appétit!

Cucumber *tsukemono*

5 slicing cucumbers ca. 1600 g (3 ½ lb)
5 g *konbu* in pieces
⅘ dl (⅓ c) cooking sake
⅖ dl (⅙ c) water
10 g (2 tsp) sea salt
5 g (1 tsp) sugar
Yuzu or lemon juice

1. Cut the cucumbers in half lengthwise. Using a spoon, scrape off the seeds and the innermost watery parts.
2. Place them in a dehydrator or oven at 40–50°C (105–125°F) for 12–24 hours. They should now weigh ca. 400 g (14 oz).
3. Cut the pieces into thin slices that are 2–3 mm (about ⅛ in) thick. Pack the cucumber and the *konbu* pieces tightly into a plastic container with a tight-fitting lid.
4. Make a marinade from the sake, water, salt, sugar, and a few drops of *yuzu* juice. One can also use the basic *tsukemono* marinade (see Flavour Accelerators).
5. Pour the marinade into the container, pressing down on the cucumber to eliminate any air pockets. Make sure the top layer is covered with the liquid. If necessary, add extra water or sake.
6. Store the container in the refrigerator. The *tsukemono* will be ready to eat in about two days or so and will keep for a couple of weeks under refrigeration.
7. For a stronger taste, substitute *dashi* (see Flavour Accelerators) for the water. To bring out other interesting nuances, add a small piece of ginger root, a clove or two of garlic, two bay leaves, or a few whole peppercorns.

> *This recipe is a variation of the simple sweet-sour cucumber salad that is common in northern European cuisine. The difference is that these become limp very quickly, whereas the tsukemono version stays crisp for a couple of weeks. It also has a much stronger cucumber taste because dehydration concentrates the aromatic substances. The pickles have sour, sweet, salty, and umami tastes and are also characterised by any added spices.*

Kohlrabi *tsukemono*

4 medium sized kohlrabies ca. 1200 g (2 ⅔ lb)
5 g *konbu* in pieces
⅘ dl (⅓ c) cooking sake
⅖ dl (⅙ c) water or *dashi* (see Flavour Accelerators)
10 g (2 tsp) sea salt
5 g (1 tsp) sugar
Yuzu or lemon juice

1. Cut each kohlrabi up into three pieces and place them in a dehydrator or oven at 40–50°C (105–125°F) for 12–24 hours, depending on their size.
2. Trim off the ends and peel the pieces, if the peel seems tough. They should now weigh 250–300 g (8–10 oz).
3. Cut the pieces into thin slices that are 2–3 mm (about ⅛ in) thick. Pack the kohlrabi and the *konbu* pieces tightly into a plastic container with a tight-fitting lid.
4. Make a marinade from the sake, water, salt, sugar, and a few drops of *yuzu* juice.
5. Pour the marinade into the container, pressing down on the slices to eliminate any air pockets. Make sure the top layer is covered with the liquid. If necessary, add extra water or sake.
6. Store the container in the refrigerator. The *tsukemono* will be ready to eat in about two days and will keep for a couple of months under refrigeration.

For a stronger taste, substitute *dashi* (see Flavour Accelerators) for the water. This recipe can also be used to pickle *daikon*. These are trimmed and sliced in half lengthwise before dehydrating, but it is not necessary to peel them at any point.

> *Kohlrabies and* daikon *are some of the best choices for making* tsukemono *that will remain extremely crisp for months. Their texture turns out to be flexible and soft, but when one gives them a hard bite the effect is so crunchy and loud that you can almost feel it reverberating in your skull. They have sweet, salty, sour, and umami tastes. Daikon pickles retain some of their delicate cabbage smell, but they are much less bitter than raw radishes and kohlrabies.*

Salt-pickled lemons

1 kg (2 lb 3 oz) organic lemons
1 chilli pepper
40 g (2 ⅔ tbsp) sea salt
15 g (1 tbsp) black peppercorns
1 l (1 qt) pickling jar

1. Rinse the lemons thoroughly, trim off the ends, cut them into wedges, and pick out the seeds.
2. Cut the chilli pepper open and remove the seeds.
3. Place a little of the salt in the bottom of the pickling jar.
4. In a bowl, squeeze together the lemons with the remaining salt, chilli pepper, and peppercorns, until a great deal of the juice comes out of the lemons.
5. Put everything into the pickling jar, leaving a bit of space at the top. Push the contents down to eliminate air pockets and squeeze out a little more juice. If necessary, add more lemon juice to cover the contents completely.
6. Allow the lemons to ferment for at least one month at room temperature. During the first week, open the jar every day to allow the gases formed by fermentation to escape, and after that only once in a while.
7. After one month, keep the lemons refrigerated. They will keep for up to a year.

Serving suggestions
These pickles can be used in most vegetable dishes to add acidity, saltiness, and fermented flavour.

The spontaneous fermentation, which is due to lactic acid bacteria, contributes umami and increases the acidity of the lemons.

5% brine: 9 ½ dl (4 c) water + 50 g (3 ⅓ tbsp) sea salt

1 kg (2 lb 3 oz) green beans

300 g (10 ½ oz) prune plums

3 sprigs of savory

3 cloves of garlic

5 g (1 tsp) black peppercorns

1 large cabbage leaf (optional)

2 l (64 oz) pickling jar

Salt-pickled beans and plums with savory

1. Heat the salt in one-third of the water until it has dissolved and then mix in the remaining cold water. Allow to cool.
2. Rinse and trim the beans, split and pit the plums, peel the garlic, and rinse the savory sprigs.
3. Put them in the pickling jar and press them down. Cover with the cooled brine, leaving about 3–4 cm (1–1 ½ inches) of space at the top.
4. Cover the top of the contents loosely with a cabbage leaf, if possible.
5. Close the pickling jar and let it stand at room temperature for a week.
6. Open the lid every day to allow the gases formed by fermentation to escape. Check the beans at this time to see if they are ready. If not, allow them to ferment for another couple of weeks.
7. Place the pickles in the refrigerator, where they can keep for quite a long time.

Serving suggestions

Serve as a garnish with other green dishes or as a small snack.

> *Lactic acid bacteria are responsible for the spontaneous fermentation and contribute sour and umami tastes to the beans. These are augmented by* koku *from the garlic, saltiness and sweetness from the plums, and aromas from the peppercorns and the savory, to create a broad palette of taste impressions.*

Savory.

Koji-marinated broccolini with chilli sauce

100–200 g (3 1/2–5 oz) broccolini

1 tbsp *shio-koji* (see Flavour Accelerators)

1 tbsp lime juice

Fish sauce, according to taste

1 tbsp sugar

Fresh chilli peppers, according to taste

Salt

1. Trim the broccolini spears and place them in a sealed plastic bag with *shio-koji*, shake to coat the vegetables, and store in the refrigerator for at least 24 hours.
2. Mix together equal quantities of lime juice, fish sauce, and sugar, as suggested above. Taste and fine-tune for sweetness, acidity, and saltiness according to personal preferences.
3. Remove the seeds from the chilli peppers and chop them finely. Depending on the size and type of peppers, add all or part of them to achieve the desired degree of spiciness. Add salt if required.
4. Refrigerate the marinade.

Serving suggestions

Arrange the broccolini spears on a plate and drizzle the marinade over them.

> *Prepared this way, the broccolini will retain most of their crispness and keep for about a week in the refrigerator. The* koji *elicits sweetness and umami, which helps to counteract the bitter taste of the vegetable. And the fish sauce contributes even more umami.*

Kimchi

1 ½ l (6 ⅓ c) water
90 g (6 tbsp) sea salt
1–1 ½ kg (ca. 2 ¼–3 ¼ lb) napa (Chinese) cabbage
200 g (7 oz) carrots
150 g (5 oz) *daikon*
10-12 spring onions
1 tbsp sesame seeds

Kimchi paste
5 g (⅕ oz) *konbu*
30 g (1 oz) peeled and trimmed garlic
30 g (1 oz) peeled ginger root
4 tsp fish sauce
4 tsp *shiro-miso* (see Flavour Accelerators)
¾ tsp chilli powder
1 tsp sugar
1 tsp sesame oil
A little water

1. To make the brine, dissolve the salt in the water.
2. Rinse the cabbage, take off the outer leaves and set them aside.
3. Massage the cabbage a little to soften it, place it in the brine, cover with the outer leaves, and let it stand at room temperature for about 6 hours.
4. For the kimchi paste: Soak the *konbu* for about an hour. Peel the garlic and the ginger and chop them coarsely. Put them together with the seaweed, fish sauce, *miso*, chilli powder, sugar, and sesame oil. Blend, adding a little water as needed to make a porridge-like consistency.
5. Remove the cabbage and the outer leaves from the brine and squeeze to remove most of the water. Place in a sieve and allow them to drain completely.
6. Peel the carrots and the *daikon* and grate them. Rinse the spring onions and cut them into strips.
7. Cut the cabbage into quarters lengthwise, trim away the cores, and slice the quarters crosswise into wide strips.
8. Place the cabbage, carrots, *daikon*, and spring onions in a bowl. Add the kimchi paste and sesame seeds. Massage together with the hands until the vegetables begin to give off a bit of liquid.
9. Scald a large pickling jar and fill with the vegetables, pressing them down so that there is a little liquid on top. Cover with the outer leaves, making sure that there is still some air space in the glass.
10. Leave the pickling jar at room temperature for 4–5 days, but open the lid each day to allow the gases formed by fermentation to escape.
11. Taste the kimchi to determine whether it is sufficiently fermented. Then store in the refrigerator where it can keep for a long time.

Serving suggestions
Kimchi is versatile and can be used as a garnish or to add taste, for example, on toast, with fried eggs, in a salad to accompany spicy dishes, or as a side dish with plant-based dishes.

Lactic acid bacteria are responsible for the spontaneous fermentation and contribute sour and umami tastes.

Koji-marinated vegetables
with herbs, Kampot pepper sauce, and fish

Serves 4

8 white asparagus spears
6–8 broccolini spears
12 large radishes
2 tbsp *shio-koji* (see Flavour Accelerators)
4 fresh fillets of fish, for example, catfish, bass, or salmon
Rye flour and butter
Herbs, including chervil, radish sprouts, and baby arugula
Kampot pepper sauce (see Flavour Accelerators) If Kampot pepper is not available, substitute with other very aromatic peppercorn

It takes at least a day to ferment the vegetables in *shio-koji*, and they taste fantastic after 3 or 4 days.

1. Peel the white asparagus and cut them into pieces 3–4 cm (ca. 1 ½ in) long. Cut up the broccolini with their stalks into elongated florets. Rinse and trim the radishes and cut them in half lengthwise.
2. Put the vegetables into a sealable bag, add the *shio-koji*, and close the bag, leaving the air inside.
3. Shake the bag so that the vegetables are well coated with *koji*, and place it in a cool place. Leave it for 1–4 days.
4. On the day of the dinner. Check the fillets for bones and debone if necessary.
5. Season the fillets with salt and pepper and coat them with the rye flour.
6. Just before serving fry the fillets in butter until they are light brown.

Serving suggestions
Brush a little Kampot pepper sauce onto the plates. Place the fried fillets on top and distribute the *koji*-marinated vegetables around them. Top with the herbs. Serve extra Kampot pepper sauce on the side.

> *Vegetables prepared cold in* koji *retain their crispness. The fermentation process elicits umami and brings out sweetness, while suppressing the bitterness that is natural in broccolini and radishes. This counteracts the very spicy taste of the sauce. Vegetarians can substitute a food with a soft texture, such as jasmine rice or soft tofu, for the fish.*

Kidney bean and *tsukemono* salad

Serves 4

250 g (9 oz) cooked kidney beans
75 g (2 ⅔ oz) *tsukemono*, from *daikon*, kohlrabies, turnips, or cucumbers
Ponzu

1. Rinse and drain the kidney beans in a colander.
2. Cut the *tsukemono* into thin slices or strips.
3. Toss the beans and *tsukemono* carefully, adding *ponzu* to taste.

Serving suggestions
This simple salad can be used as a small appetiser or side dish.

> *The addition of the crispy and crunchy tsukemono to the cooked kidney beans that can have a mealy and dry mouthfeel makes the sensory perception of the mixture much more than the sum of the two parts. This general principle is called texture contrast.*

Koji-marinated vegetables with herbs, Kampot pepper sauce, and fish.

Celeriac salad
with lemons, apples, golden raisins, mustard seeds, and celery seeds

Serves 4

1 tbsp mustard seeds
2 juicy lemons
Ca. 800 g (1 lb 12 oz) celeriac
3 tart apples
100 g (3 ½ oz) golden raisins
A little salt
Fresh herbs or celery tops

Dressing
1 ½ dl (3/5 c) white wine vinegar
1 tbsp sugar
½ dl (1/5 c) water
1 tsp celery seeds
20 g (4 tsp) Dijon mustard
¼ tsp salt
2 tsp nutritional yeast flakes
1 anchovy fillet, optional

1. Put all the dressing ingredients in a blender and blend thoroughly.
2. Soak the mustard seeds in water and squeeze the juice from the lemons.
3. Peel the celeriac. Cut it first into slices and then julienne them.
4. Toss the celeriac strips with the lemon juice and season to taste with salt.
5. Grate the apples and mix them with the celeriac. Drain the mustard seeds and add to the mix together with the raisins.
6. With a large kitchen spoon, work the dressing into the salad, so that the celeriac becomes a little pliable. Fine-tune the taste with a little salt. Set aside for about 30 minutes before serving. Garnish with fresh herbs or delicate celery tops.

Serving suggestions
The celeriac salad pairs well with a small luncheon dish. Optionally, adding a few tamarind cashew nuts (see Flavour Accelerators) or salted, smoked peanuts (see Flavour Accelerators) introduces an additional pleasing, crunchy mouthfeel.

Because the celeriac is julienned it has a crisp mouthfeel and seems to lose some of its celery taste. Apples and raisins contribute sweetness, the lemon juice acidity, and the mustard seeds some spiciness. The dressing adds umami from both the anchovy fillet and the nutritional yeast, as well as sweetness, salt, and a sharp taste from the mustard. All five basic tastes and trigeminal sensory perception are represented in this dish.

Radicchio
with dried olives, balsamic vinegar, and cheese

Serves 4

20 pitted Kalamata olives
1 firm torpedo-shaped red radicchio
40 g (1 ⅓ oz) pistachios
100 g (3 ½ oz) creamy goat cheese
4 tbsp aged balsamic vinegar or *vincotto*
Olive oil
Tender beet leaves, wood sorrel, or arugula, for decoration
Salt flakes

1. Place the Kalamata olives in a dehydrator or oven at 50°C (125°F) for 3 hours to dehydrate them.
2. Wrap the radicchio in plastic film so that it is easier to slice it.
3. Toast the pistachios lightly on a dry skillet.
4. Spread a generous spoonful of the cheese in the middle of four plates.
5. Cut the radicchio into quarters lengthwise. Remove the plastic and place them on top of the cheese.
6. Optionally, a culinary torch can be used to char the radicchio a little to add a slightly burnt taste.

Serving suggestions
Distribute the remaining cheese in dollops on top of the radicchio. Crush the dried olives and the pistachios with the blade of a knife and sprinkle them on top. Drizzle with a little balsamic vinegar and olive oil. Decorate with a few leaves and finish with a sprinkling of salt flakes.

Balsamic vinegar contributes acidity and an intense sweetness that pairs with the creaminess of the goat cheese to take the edge off the bitter taste of the radicchio.

Wakame salad with tofu

Serves 4

5 g (¼ oz) dried *wakame*, cut into small pieces
250 g (8 ⅔ oz) coarsely grated carrots
1 English cucumber, cut into cubes
250 g (8 ⅔ oz) firm tofu, cubed
Salt
Toasted sesame seeds, optional

Dressing
2 tbsp rice vinegar
2 tbsp soy sauce
2 tbsp marinade from pickled ginger (*gari*)
A little finely chopped pickled ginger, optional (see Flavour Accelerators)

1. Soak the *wakame* in warm water for 10 minutes and then drain.
2. Mix the ingredients to make the dressing.
3. Mix together the seaweed, carrots, cucumber, and tofu. Stir in the dressing.
4. Season with salt. Optionally sprinkle toasted sesame seeds on top.

Serving suggestions
This salad can be served as an appetiser or as a side with fish dishes.

This salad is fresh and juicy and has umami from the soy sauce and rice vinegar. The pickled ginger adds sharp, acidic tastes to the vegetables.

Radicchio with dried olives, balsamic vinegar, and cheese.

Cucumbers with sesame seeds, *ponzu*, chilli, and ginger

Serves 4

2 organic English cucumbers
1 tbsp sesame seeds
50 g (1 ⅔ oz) pickled ginger (*gari*) (see Flavour Accelerators)

Sauce
50 g (1 ⅔ oz) shallots
Small chilli pepper, about 7 g (¼ oz)
1 dl (⅖ c) Japanese soy sauce
1 dl (⅖ c) *ponzu*
10 g (2 tsp) tahini paste or *goma-shio* (see Flavour Accelerators)
Chilli flakes, to taste

1. Trim the ends of the cucumbers and cut each of them into 3 pieces, approximately 8 cm (3 in) long. Next cut each of the pieces in half lengthwise and then each half into three equal segments.
2. Trim the seeds away from the segments and try to keep the cut surfaces evenly flat.
3. For the sauce: Chop the shallots and the chilli pepper very finely. Place them in a saucepan together with the other ingredients and bring them to a boil. Cook for ca. 10 minutes until the sauce thickens and the liquid has reduced by about a half.
4. Taste the sauce and add a few chilli flakes if needed.
5. Toast the sesame seeds on a dry skillet until they are light brown.

Serving suggestions
Arrange 6 pieces of cucumber per person, two by two in three layers with the sauce in between each layer and extra on top. Sprinkle with the sesame seeds and place the pickled ginger on the side.

> *In this recipe, a cucumber becomes more interesting and plays a role similar to that of a fish fillet. The thickened sauce adds umami to the crisp cucumber, while the intense taste of the* ponzu *provides a contrast to the fresh, biting, strong taste from the chilli and pickled ginger. The oil in the sesame seeds releases a slightly pungent, nut-like aroma.*

Miso soup with seaweed

Serves 4

1 l (4 c) *dashi* (see Flavour Accelerators)

Small pieces of *wakame* or winged kelp

2 tbsp *miso* paste

Other ingredients, for example, tofu, mushrooms, green onions, shellfish, fish (optional)

1. Heat the *dashi* and add the seaweed to it, but do not allow it to boil.
2. Optionally, add other ingredients, such as small cubes of tofu, thin slices of mushrooms, green onion rings, a shrimp or two, according to how long it would take to warm them through sufficiently. Bits and pieces of leftover vegetables are also suitable.
3. Dilute the *miso* paste with lukewarm water to make it runny and stir it into the soup.
4. Heat the soup to just below the boiling point and serve immediately.

Serving suggestions

The soup can be served as a small appetiser with just a few green onion rings, almost as a decoration, or turned into something more substantial with pieces of fish, shellfish, mushrooms, and tofu.

> *Miso soup provides a daily infusion of umami in Japanese cuisine, starting even as early as breakfast. It is fast and easy to prepare if dashi and miso paste are on hand.*

Pea soup with scallops

Serves 4

2 large onions

3 cloves of garlic

4 dl (1 ³⁄₅ c) basic vegetable stock (see Flavour Accelerators)

1 kg (2 lb 3 oz) small shelled fresh or frozen peas

2 dl (⁴⁄₅ c) milk

Neutral oil

Salt, seaweed salt, pepper

4 large scallops

Dried sugar kelp, winged kelp, or dulse flakes

1. Peel and chop the onions and the garlic coarsely. Cook them in a pot in a little oil until they are transparent.
2. Add the stock and cook for ca. 5 minutes.
3. Set aside 3 tablespoons of the peas for later use.
4. Add the rest of the peas and the milk to the stock and reheat to cook the peas.
5. Blend the contents of the pot, adding a little more milk if necessary to get the desired consistency. Season with salt and pepper.
6. Mix in the peas that were set aside and reheat.
7. Sauté the scallops quickly on both sides on a dry, warm skillet.

Serving suggestions

Serve the soup in individual bowls with a scallop added to each. Sprinkle a few dried seaweed flakes on top.

> *This recipe is designed to take advantage of our knowledge about synergistic umami. In this case the very intense umami taste is derived from the interaction between the free glutamate in the peas and the free adenylate in the scallops. It is intensified by using the vegetable stock, which is another source of umami. There is also a pleasing sweetness from the peas and the scallops. In addition, the sensation of koku comes from the garlic.*

Pea soup with scallops.

Onion soup—the real thing!
with baked cheese dumplings

Serves 4

1 kg (2 lb 3 oz) yellow onions
2 cloves of garlic
5–6 sprigs of thyme
2 tbsp olive oil
Nutmeg according to taste
A pinch of cayenne powder
1 ½ l (6 ⅓ c) liquid, for example, vegetable or chicken bouillon, potato cooking water, or water

Cheese balls

200 g (7 oz) very firm, strong cheese, for example, Gruyère
25 g (1 oz) Parmesan cheese
100 g (3 ½ oz) fresh, soft cheese, possibly creamy goat cheese
1 egg
10 g (2 tsp) breadcrumbs
¼ tsp cayenne powder
1 ½ dl (6 ⅓ c) neutral oil for deep-frying
Salt

Onion soup
1. Trim the onions, cut them in half lengthwise, and then slice them. Peel the garlic cloves and crush them with the heel of your hand. Rinse the thyme sprigs.
2. Warm the olive oil in a saucepan with a thick bottom and add the garlic, thyme, and cayenne powder.
3. Add the sliced onions and grate a little nutmeg over them. Let everything simmer covered on low heat, stirring occasionally. Then remove the lid and let it continue to simmer for at least 45 minutes. Even longer is better. Pick out and discard the garlic cloves and the thyme sprigs.
4. Add the liquid and let the soup continue to simmer for another 25 minutes. Season to taste.

Cheese balls
1. Grate the Gruyère and the Parmesan cheese finely.
2. Mix together the three types of cheese, egg, breadcrumbs, and cayenne powder and season with salt.
3. Form the mixture into small, even balls, about the size of a walnut.
4. Heat the oil to 150°C (300°F) and deep-fry the balls working in batches to avoid crowding them. They will usually turn themselves in the oil. When they are golden, remove them with a slotted spoon and place them on absorbent paper towels.

Serving suggestions

Warm the cheese balls in the oven at 80°C (175°F) for 3–4 minutes. Serve the soup piping hot straight from the pot with the warm cheese balls on the side.

> *Taking the time to simmer the onions over a long period of time is the end all and be all in determining how the soup will taste. The onions caramelise and develop a very rich taste and strong koku sensation. Potato cooking water and bouillon add umami.*

Sautéed spinach, mung beans, and wood ear mushrooms
with nutmeg omelette and Greek yogurt

Serves 4

200 g (7 oz) mung beans
500 g (18 oz) spinach, preferably coarse-leaved
4 eggs
Butter
Salt flakes
Whole nutmeg
400 g (14 oz) mushrooms, for example, wood ears, *shiitake*, beech mushrooms
100 g (3 ½ oz) shallots
1 dl (2/5 c) soy sauce
1 tbsp oyster sauce (optional)
Finely milled salt
Crisp garlic (see p. 268)

Dressing
1 ½ dl (3/5 c) Greek yogurt
½ bunch of tarragon
1 clove garlic
1 organic lemon
Salt
Pinch cayenne pepper

1. Spoon the yogurt into a bowl. Chop the tarragon coarsely, mince the garlic, and grate the lemon peel.
2. Stir into the yogurt and mix it all with a hand blender. Season with salt and cayenne pepper. Set aside.

Omelette
1. Rinse the mung beans, cook them in lightly salted water for ca. 30 minutes, and then drain them.
2. Rinse the spinach, spin to remove excess water, and trim off any coarse stalks. Place the spinach on absorbent paper towels and set aside.
3. Crack the eggs into a bowl and stir them together a little without beating them.
4. Melt a little butter in a 28 cm (ca. 12 in) skillet on low heat. Pour in the egg mixture like a pancake.
5. Cook the eggs on low heat, sprinkle with the salt, and grate the nutmeg over the entire surface when the eggs begin to set.
6. When it has cooked through completely, turn the 'pancake' out onto a cutting board. Cut it up into thin strips.
7. Clean the mushrooms and cut or tear them into uneven pieces.
8. Finely chop the shallots.
9. Sauté the mushrooms in a little butter together with the chopped shallots in a skillet on medium heat. Add the cooked mung beans and stir a couple of times. Add the soy sauce and the oyster sauce, stir again, and then transfer to a bowl.
10. Wash the skillet. Divide the spinach into 2 or 3 batches and sauté in a little butter on medium heat. Season with finely ground salt.

Serving suggestions
Arrange the warm beans, mushrooms, and spinach on a large platter. Place the egg strips on top and sprinkle with the crisp garlic. Serve the dressing on the side.

> *If fresh wood ear mushrooms are not available, substitute dried ones. These fungi have an interesting texture, but very little taste. This recipe is designed to showcase the spinach in a wonderful combination with the gentle texture of the mung beans and the tough, almost elastic, mouthfeel of the wood ear mushrooms. The taste experience is enhanced by an abundance of umami from soy sauce, oyster sauce, and other mushrooms, all of it enlivened by that 'aha moment' provided by the nutmeg seasoned eggs and the crisp garlic.*

Rizza

Serves 1

1 egg
A little salt and pepper
Chopped herbs or spices
Leftovers
1 sheet of rice paper

If the leftovers are bits of meat or fish, they should be cooked through before use.

1. Beat the egg and season with salt and pepper.
2. Chop up the leftovers into small pieces.
3. Moisten one side of the rice paper with water and place it damp side down on an ungreased skillet.
4. Cover the rice paper with the beaten egg.
5. Spread the chopped leftovers on top. Sprinkle with chopped herbs or spices to taste.
6. Cook at medium heat until the egg has set. Fold the *rizza* in half and enjoy.

This ingenious rice paper 'pizza' is an innovative way to involve children and young people in having fun while preparing food. While rizza *mimics the fast food of which they are fond, it is a truly nutritious and tasty small meal. It uses up leftovers of all kinds creatively and makes allowance for individual preferences.*

We came up with the idea of rizza at Taste for Life as an outreach project to school children. The aim was to find a recipe that the students could use to make a quick little after-school snack that could compete with potato chips, candy bars, and so on. This was actually inspired by a trip to Vietnam where, after many fantastic and varied experiences with street food, we came across the ultimate version turned out by a young man who had set up the world's simplest restaurant. It consisted of two flatbed toasters, a pile of rice paper, a tray of eggs, and different fillings such as dried fish, banana blossoms, mango, and broccoli. We were served a snack that was totally crispy, rich in umami and other good tastes, prepared by one of the most fearlessly ambitious cooks we have ever met. It was absolutely brilliant.

Rizza is not just for young people, and we urge everyone to try making this simple dish with whatever tasty leftovers are on hand in the refrigerator. But readers might also seize the chance to turn making rizza *into a fun family experience by involving children or grandchildren—it is almost guaranteed to be an occasion for a good chat as well.*

Kale pancakes
with runny eggs

Serves 4

Pancakes
- 20 g (4 tsp) melted, cooled butter
- 1 egg
- 1¼ dl (½ c) milk
- ½ tsp salt
- 45 g (3 tbsp) flour
- A little extra butter

Filling
- 200 g (3½ oz) kale
- 20 g (1⅓ tbsp) butter
- 20 g (1⅓ tbsp) flour
- 2 dl (⅘ c) milk
- 30 g (2 tbsp) Parmesan cheese, grated
- Ground nutmeg
- Salt and pepper
- 1 egg

Soft boiled eggs
- 4 small to medium organic eggs

Small salad
- Herbs, for example, fennel tops, tarragon, sprouts
- Salt and pepper

4 small aluminium tart pans, ca. 7 cm (3 in) in diameter and 4 cm (1½ in) high

Eggs are the central feature of this dish, becoming part of the sauce and adding creaminess. The egg yolks and the Parmesan cheese contribute umami.

Pancakes
1. Melt the butter and let it cool to room temperature.
2. Whisk together the egg, milk, and salt. Stir in the flour and then the butter.
3. Divide into four portions and cook four pancakes in a little butter in a ca. 24 cm (ca. 10 in) skillet on medium heat. Set aside.

Filling
1. Rinse and quickly blanch the kale in boiling water; drain and plunge it into very cold water with a few ice cubes.
2. Spin to remove the water and chop the kale finely.
3. Melt the butter in a small pot, over low heat. Whisk in the flour until it is smooth. Gradually whisk in the milk, bring to a boil, immediately lower the heat, and continue to stir the sauce until the taste of the flour has disappeared and it has thickened. Stir in the grated Parmesan cheese. Season with a little ground nutmeg, salt, and pepper.
4. Stir the chopped kale into the sauce. If a smoother consistency is desired, some of the kale can first be put through a parsley mill or pulsed in a blender.
5. Let the sauce cool and then whisk in the egg.

Soft boiled eggs
1. Put the eggs in a pot with cold water, bring to a boil, and cook the eggs for ca. 4 minutes, depending on their size. It is important that the eggs should remain very soft.
2. Cool them in cold water and then peel carefully.

Filled pancakes
1. Preheat the oven to 185°C (365°F).
2. Place the pancakes on a cutting board. Cut them in half and trim to make a 7 cm (3 in) strip from each half. Eight strips in all.
3. Grease the small tart tins and place two strips in each to form a cross. Gently push the pancake pieces against the sides of the tins to cover them, but allowing some of the strips to hang over the edge.
4. Put a spoonful of the kale sauce and an egg in each tin. Fill the tins with sauce to cover the egg. Then cover the top with the parts of the strips that were hanging over the edge.
5. Bake for ca. 15 minutes.
6. Make a small green salad with herbs such as fennel tops, tarragon, and sprouts. Season with salt and pepper.

Serving suggestions
Carefully remove the pancakes from the tins, plate them, and sprinkle the greens on top.

Tortilla and figs
with goat cheese and potatoes

Serves 4

500 g (1 lb 1 ½ oz) firm potatoes

1 medium sized onion

4 fresh ripe figs

1 dl (2/5 c) olive oil

Black pepper and salt

6 large organic eggs

2 or 3 fig leaves, if available

100 g (3 ½ oz) firm goat cheese at room temperature broken up into smaller pieces

1 stalk of fresh rosemary

1. Peel the potatoes, cut them up into cubes, ca. 2 × 2 cm (¾ × ¾ in). Rinse them thoroughly and dry them off.
2. Peel and chop the onion.
3. Cut the figs up into quarters.
4. Pour the oil into a skillet and fry the potatoes, stirring them carefully. Add the chopped onion when the potatoes are almost done. Let the mixture continue to fry until the potatoes are light brown. Season with salt and remove from the heat.
5. Whisk together the eggs in a bowl, season with salt and freshly ground pepper. Stir in the potato-onion mixture.
6. If available, place the fig leaves in an oven-proof dish. Otherwise grease the dish. Pour in the egg mixture and distribute the fig quarters and small pieces of the goat cheese in it. Top with the rosemary stalk.
7. Bake the tortilla in a preheated oven at 180°C (355°F) for ca. 30 minutes.

Serving suggestions
Serve directly from the baking dish, possibly with some slices of rye bread and salad, for example, Celeriac salad (p. 182).

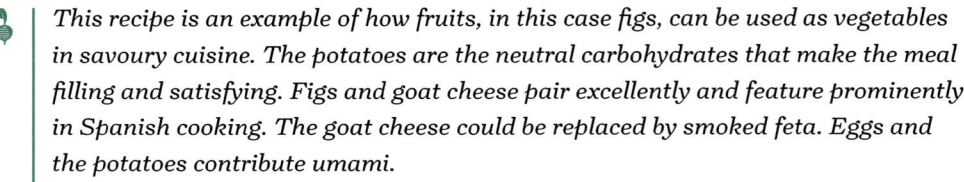

This recipe is an example of how fruits, in this case figs, can be used as vegetables in savoury cuisine. The potatoes are the neutral carbohydrates that make the meal filling and satisfying. Figs and goat cheese pair excellently and feature prominently in Spanish cooking. The goat cheese could be replaced by smoked feta. Eggs and the potatoes contribute umami.

Serves 4

1 celeriac, ca. 800 g (1 ¾ lb)
4 organic eggs
Butter for frying
Ca. 50 g (3 tbsp) grated Parmesan cheese, very old cheddar, or other hard cheese
Sprigs of summer savory or thyme
Salt flakes and pepper
1 bunch of chervil or flat leaf parsley, finely chopped

Sprouted grains
100 g (3 ½ oz) wheat berries
Water
Lemon juice

***Miso*-mayonnaise**
2 egg yolks
1 tsp *shiro-miso* (see Flavour Accelerators)
1 tsp mustard
2 tbsp apple cider vinegar
Salt
2 dl (⅘ c) neutral oil

Fried eggs on a bed of celeriac
with dry cheese, *miso*-mayonnaise, and sprouted grains

Sprouted grains
1. Soak the wheat berries in water for 24 hours. Discard the water and place the wheat berries on one half of a clean, very damp dish towel. Cover the berries with the other half of the towel.
2. Place the towel in a cool place, but do not refrigerate. Rinse the sprouts once every day in fresh water with a little lemon juice for 4 days. Be careful, especially at first, so that the sprouts are not broken off.

Miso-mayonnaise
1. All the ingredients for the mayonnaise should be at room temperature.
2. In a deep bowl, whisk together the egg yolks, *miso*, mustard, vinegar, and salt. Whisk in the oil slowly, at first drop by drop, and then by the teaspoonful. Stop when the mayonnaise has the desired consistency.

Celeriac and fried eggs
1. Peel and thoroughly rinse the celeriac and then cut it up into 4 slices, 1 ½ cm (⅔ in) thick. Place them on a baking sheet and bake them with the savory or thyme at 200°C (390°F) for ca. 15 minutes until they are tender, but still a little firm.
2. Fry 4 eggs in butter on medium heat.

Serving suggestions
Spread *miso*-mayonnaise on the celeriac slices and place a fried egg on each. Sprinkle with the salt flakes, freshly ground pepper, and a generous amount of grated cheese. Finally top with the chopped herbs and sprouted wheat berries. This dish can be paired with beans, for example, fermented beans, bean salad, or polenta fritters (p. 256) to make up a very filling main course.

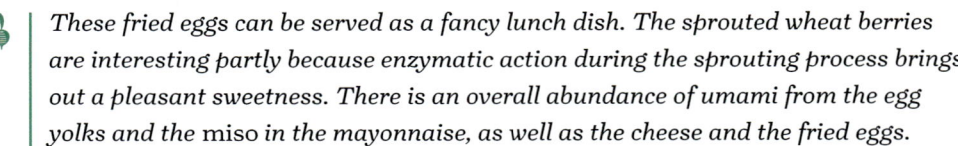

These fried eggs can be served as a fancy lunch dish. The sprouted wheat berries are interesting partly because enzymatic action during the sprouting process brings out a pleasant sweetness. There is an overall abundance of umami from the egg yolks and the miso *in the mayonnaise, as well as the cheese and the fried eggs.*

'Greens to go' casserole
with button mushrooms, celery, spinach, and horseradish pesto

Serves 4 as an entrée or 8 as an appetiser or side dish.

350 g (12 oz) assorted vegetable odds and ends ('greens to go')

200 g (7 oz) button mushrooms

100 g (3 ½ oz) celery

100 g (3 ½ oz) spinach

2 dl (⅘ c) vegetable bouillon or potato cooking water (see Flavour Accelerators)

50 g (3 ⅓ tbsp) butter

50 g (3 ⅓ tbsp) flour

1 ½ dl (⅗ c) milk

5 large eggs, separated into whites and yolks

60 g (4 tbsp) finely grated Parmesan cheese

Salt and pepper

Butter to grease the casserole dish

½ dl (⅕ c) crisp topping, for example, soy toasted pumpkin seeds, almonds, or hazelnuts, or breadcrumbs (see Flavour Accelerators)

Horseradish-parsley pesto with capers (see Flavour Accelerators)

1. Cut the vegetable odds and ends into smallish pieces, so that one can still tell them apart and feel their different textures when one eats them.
2. Rinse the button mushrooms and the celery; cut the mushrooms up into small pieces and the celery into pieces ½ cm (⅕ in) thick.
3. Rinse the spinach, chop it coarsely, and mix with the bouillon or potato cooking water.
4. Prepare a thickener: Melt the butter, whisk in the flour, add the milk, and warm it until it thickens.
5. Add the spinach and bouillon (or potato cooking water) and bring it to a boil. Allow it to cool a little until the egg yolks can be added without coagulating.
6. Whisk the egg yolks, add them to the sauce, and stir until the mixture is uniform.
7. Carefully stir the vegetable odds and ends, celery, mushroom pieces, and ⅔ of the cheese into the mixture.
8. Season with salt and pepper.
9. Beat the egg whites until they form stiff peaks; fold them carefully into the mixture.
10. Grease a casserole dish of an appropriate size, about 2 litres (2 quarts), and sprinkle the crisp topping on the sides.
11. Pour the vegetable mixture into the casserole dish. It should not come all the way up to the top of the sides. Sprinkle the rest of the cheese on top.
12. Bake the casserole in a preheated oven at 175°C (350°F) for 60–70 minutes.
13. The mixture can also be baked in small individual dishes and used as an appetiser or side dish. The baking time is then reduced to ca. 20 minutes.

Serving suggestions

Serve the casserole piping hot as an appetiser, side dish, or entrée with the horseradish pesto as a condiment.

A casserole can have a disappointing mouthfeel and taste, but if one introduces a variety of textures, for example, from mushrooms and the different sizes of the vegetable pieces, it becomes more interesting. The horseradish pesto adds the appealing taste of the herbs, acidity from the capers, and heat from the horseradish. Exchanging the breadcrumbs often used as toppings on casseroles for the crushed soy toasted pumpkin seeds contributes umami in symbiosis with the Parmesan cheese, eggs, and mushrooms.

Green fricassee

with pineapple and Kampot pepper sauce, Cambodian style

Serves 4

500 g (1 lb 1 ½ oz) ripe pineapple pieces, after peeling

300 g (10 ½ oz) carrots, preferably of different colours

150 g (5 ⅓ oz) Swiss chard or celery

150 g (5 ⅓ oz) spring onions

200 g (7 oz) onions

1 organic lime

2 tbsp neutral tasting oil

150 g (5 ⅓ oz) green beans or green bell peppers

Salt flakes

Kampot pepper sauce (double portion) (see Flavour Accelerators)

Cooked jasmine rice

1. Peel the pineapple and cut it up lengthwise into pieces that are ca. 2 cm (1 in) thick.
2. Then cut the pieces into thin strips.
3. Peel and rinse the carrots and the Swiss chard and cut them up into strips. Trim and cut the spring onions into strips. Peel and trim the onions and cut them into segments.
4. Cut the lime in half and cut each half into three segments.
5. Place a little oil in a skillet and add the vegetables, pineapple, and lime. Sauté at a low to medium heat until done and slightly toasted.
6. Trim the end of the beans or remove the seeds from the green peppers and cut them into strips. Blanch them in boiling water, and add them to the vegetables in the skillet.
7. Season with salt flakes.

Serving suggestions

This dish can be served as is in the skillet or arranged on a platter. The most important element is the Kampot pepper sauce that is served on the side. Rice is an essential accompaniment to the vegetable fricassee.

This dish is a meatless fricassee. Fruit, in this case pineapple, is the 'meat' in the dish, where it also acts as a vegetable. It is the special Kampot pepper sauce that gives this dish a lift and evokes the tastes of Cambodian cuisine. Likewise, it makes one want to eat a lot of vegetables.

Sicilian ratatouille

Serves 4 to 6

400 g (14 ½ oz) aubergines
Virgin olive oil
2 tbsp capers, soaked in water
200 g (7 oz) zucchini, 1 medium sized
2–4 artichoke hearts
5 stalks of celery
200 g (7 oz) shallots
2 cloves of garlic
400 g (14 ½ oz) very ripe tomatoes
2 tbsp tomato paste
Fresh oregano, chopped
White wine vinegar
Sugar, salt, and pepper
2 tbsp pistachios

1. Cut the aubergines into small cubes. Salt them and place them in a sieve to draw out some of their moisture for ca. 30 minutes. Then rinse them quickly and dry them.
2. Soak the capers in cold water for ca. 15 minutes and then drain off the water.
3. Toast the aubergine cubes in olive oil in a pot until they are light brown. Remove them with a slotted spoon and set aside.
4. Peel the zucchini and cut it as well as the artichoke hearts into cubes and the celery into slices. If fresh artichokes are not available, substitute ones preserved in oil in a jar.
5. Trim and peel the shallots and garlic and chop them finely.
6. Put the zucchini, artichoke, celery, onion, and garlic pieces into the pot and sauté them lightly. Chop the tomatoes coarsely and then add them together with the tomato paste. Simmer for 10 minutes, covered.
7. Stir in the aubergine cubes, capers, and oregano. Season with wine vinegar, sugar, salt, and pepper.
8. Toast the pistachios in a dry skillet.

Serving suggestions

Serve in individual bowls or ladle directly from the pot. Sprinkle the pistachios on top. The ratatouille will taste even better if it has been kept in the refrigerator and re-heated the next day.

 This dish is a Sicilian-inspired version of ratatouille, which is enriched with umami from the ripe tomatoes, tomato paste, and capers. Onions and garlic elicit a koku *sensation. This type of ratatouille is lighter than the traditional Niçoise version.*

Mashed vegetables with rutabaga 'bacon'
and a topping of flax seeds, seaweeds, and nutritional yeast flakes

Serves 4—6

Coarse mash
Ca. 1 kg (2 lbs 4 oz) rutabagas
300 g (2/3 lb) carrots
600 g (1 1/3 lb) baking potatoes
3 or 4 spring onions
2 laurel leaves
3 sprigs of rosemary
50 g (3 1/3 tbsp) butter
Salt and pepper

Rutabaga 'bacon'
3 sprigs of rosemary
1 clove of garlic
Peel from 1/2 organic lemon
3 slices rutabaga, 1 cm (2/5 in) thick, ca. 250 g (9 oz)
Smoking chips and rosemary stalks
1 tbsp olive oil
Salt and pepper

Cream sauce
40 g (1 1/3 oz) horseradish
1 1/2 dl (2/3 c) crème fraîche or aspic
Salt and pepper
Hemp seed crunch (see Flavour Accelerators)

For smoking
A smoker with a grill inside

This recipe features rutabagas, old-fashioned and often overlooked vegetables, in a dish with an abundance of umami and an extra infusion of aroma thanks to the smoked 'bacon.' Alternatively, one can make the dish with celeriac instead of rutabagas.

Vegetables
1. Peel the rutabagas, carrots, and potatoes.
2. Slice the whites of the spring onions into fine rings and set aside.
3. Cut the rutabagas crosswise into slices about 1 cm (2/5 in) thick. Set aside about 250 g (1/2 lb).
4. Cut the remaining rutabagas, carrots, and potatoes up into 2–3 cm (ca. 1 in) chunks. Place them in a pot, cover with water, and add the laurel leaves and rosemary sprigs.

Smoking and preparation of rutabaga 'bacon'
1. Pull the leaves from the stems of the rosemary sprigs and chop them finely.
2. Put the garlic through a press, grate the lemon peel, and set aside.
3. Place the smoking chips in the bottom of the smoker and turn it on.
4. Place the rutabaga slices on the grill and smoke them at low heat for 6–8 minutes. Transfer them to a cutting board.
5. Cut the slices up into 1 cm (2/5 in) cubes and put them in a small oven-proof skillet with the oil.
6. Toss them with the chopped rosemary, garlic, and lemon peel. Season with salt and pepper.

Cream sauce
1. Grate the horseradish, stir it into the crème fraîche, and season with salt and pepper. If fresh horseradish is not available, substitute a very small amount of already prepared horseradish.

Vegetable mash
1. Preheat the oven to 225°C (440°F).
2. Cook the already peeled rutabagas, carrots, and potatoes in water, until they are very soft, about 25–30 minutes.
3. Bake the rutabaga cubes for ca. 8 minutes, until they are done, but still a little firm.
4. Drain the vegetables and place them back on the element briefly to steam off the last bit of moisture. Put half the vegetables in a serving dish and cover to keep warm.
5. Thoroughly mash the remaining vegetables with the butter. Mix with the remaining vegetable cubes. Season with salt and pepper.

Serving suggestions
Put the warm rutabaga mash in a bowl and sprinkle the rutabaga 'bacon' and spring onion rings on top. Serve the cream sauce and hemp seed crunch in separate bowls. Seaweed granules, *za'atar*, or *goma-shio* (see Flavour Accelerators) can be substituted for the hemp seed crunch.

Aubergines, preferably long, thin ones

Neutral oil

Miso (any kind)

Tahini

Deep-fried aubergines (*nasu dengaku*) with *miso*

1. Trim the top of the aubergines. Cut them in half lengthwise and deep-fry them until they are cooked through.
2. Mix equal portions of *miso* and tahini. If necessary, add a little water to make the paste spreadable.
3. Score a diamond pattern lightly in the cut side of the aubergines and spread an even layer of the *miso*-tahini paste on them.
4. Grill the aubergines until their surfaces are golden and a little crisp.
5. The recipe can be varied as follows, but then is no longer vegan: Anchovy paste, possibly with a little pressed garlic, may be substituted for the *miso*. Panko crumbs and nutritional yeast flakes are sprinkled on top before the aubergines are grilled.

Serving suggestions

Serve the aubergines as they are directly from the grill. To eat them, dig out the aubergine flesh with a spoon.

> Nasu dengaku *is a classical dish from the Japanese vegan monastery cuisine* (shōjin ryōri). *The aubergines take on a meaty texture and the miso provides ample umami.*

Juicy aubergines
with ginger and tomatoes

Serves 4

2 small, firm aubergines, ca. 600 g (1 lb 5 oz)

20 g (4 tsp) sea salt

300–400 g (2/3–1 lb) small ripe tomatoes, preferably in different colours and sizes

100 g (3 ½ oz) red onions

1 bunch spring onions, ca. 100 g (3 ½ oz)

100 g (3 ½ oz) celery

75 g (2 ½ oz) ginger root

80 g (2 2/3 oz) bacon (optional)

2 tbsp olive oil

1 l (4 c) vegetable bouillon (see Flavour Accelerators)

Fresh basil

1. Trim the top of the aubergines. Cut them up into lengthwise slices, ca. ½ cm (1/5 in). There should 3–4 slices per person.
2. Score the slices on one side and rub sea salt into them. Allow them to drain for about one hour. Rinse the slices and dry them.
3. Cut a slit in the bottom of the tomatoes, blanch them in boiling water, remove them and plunge them into ice-cold water. Peel them.
4. Trim and peel the red onions. Trim the spring onions and celery. Cut the onions in half and then in small segments. Slice the spring onions and celery, setting aside the small green tops of the celery, if there are any.
5. Peel and mince the ginger root very finely.
6. Cut the bacon up into small pieces.
7. Pour olive oil into a skillet and sauté the aubergine slices a few at a time on both sides until they are light brown. Place them on absorbent paper towels.
8. Pour the vegetable bouillon into a saucepan, together with the bacon (if using), whole tomatoes, red onion, celery, spring onion, and ginger. Allow the mixture to simmer for 15 minutes and season to taste with salt.

Serving suggestions
Roll up the aubergine slices and place them in individual pasta bowls. Divide the bouillon and the solids among the bowls. Garnish with fresh basil leaves and the celery tops. Serve with jasmine rice or rustic bread on the side.

Even though it might resemble a soup, this is a juicy, meaty dish served in a bouillon. The resemblance to meat comes from the way the aubergines are prepared, which leaves them with a slightly tough texture and mouthfeel. The actual meat-like taste is due to the abundant umami in the ripe tomatoes. The bacon adds another range of tastes and some smoky aromas. This dish is very 'moreish.'

Serves 4

2–3 firm aubergines, ca. 800 g (1 ¾ lb)

Sea salt

60 g (2 oz) dried *shiitake*

500 g (1 lb 1 ½ oz) small red beets

A little apple cider vinegar and salt

2 dl (⅘ c) olive oil for deep-drying

25 g (5 tsp) sesame seeds

2 tbsp red *miso*

300 g (10 ½ oz) fresh spinach

Fruit vinegar marinade (see Flavour Accelerators)

A sprig of thyme

Butter

Aubergines *au gratin*
with sesame seeds, *miso*, red beets, and *shiitake*

1. Trim the top of the aubergines and cut them into slices ca. 2 ½ cm (1 in) thick. Score one side of the slices in a diamond pattern about ½ cm (¼ in) deep.
2. Salt the slices using 35 g (2 ½ tbsp) per 1 kg (2 ¼ lb) of aubergines. Let them stand on a wire rack for about an hour.
3. Rinse the aubergine slices and dry them.
4. Soak the *shiitake* in lukewarm water for ca. 45 minutes. Slice them into strips.
5. Cook the beets for 30–40 minutes in lightly salted water until they are done, but still firm. Drain the cooking water and place the pot under cold running water until the peels are easy to slip off, but the beets are still warm.
6. Remove the peels and cut the beets up into segments about 2–3 cm (¾–1 ¼ in) thick. Place them in a bowl and season with salt and apple cider vinegar.
7. Warm up the olive oil in a large skillet and fry the aubergine slices a few at a time on both sides. When done, place them on paper towels to absorb excess oil.
8. Toast the sesame seeds on a small dry skillet until they are golden. Mash them in a mortar or grind them in a spice mill and mix them with the *miso* to make a paste.
9. Cover the scored sides of the aubergines with the paste.

Just before serving:

1. Rinse the spinach and tear it into smaller pieces. Set aside.
2. Grill the aubergine slices in the oven at 185°C (365°F) for 5–6 minutes.
3. Warm the beets in the fruit vinegar marinade and season with salt and pepper.
4. Fry the *shiitake* strips in butter with the thyme sprig. Season with salt and pepper.
5. Wilt the spinach very quickly in a skillet with a little butter and 1 tbsp water. Season and drain in a colander.

Serving suggestions

Arrange the warm beet pieces on individual plates or a platter and cover with the aubergine slices. Top with the *shiitake* strips and add the spinach around the sides.

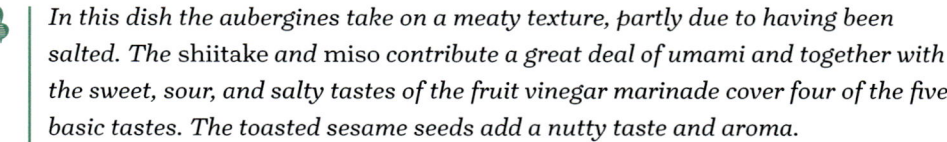

In this dish the aubergines take on a meaty texture, partly due to having been salted. The shiitake *and* miso *contribute a great deal of umami and together with the sweet, sour, and salty tastes of the fruit vinegar marinade cover four of the five basic tastes. The toasted sesame seeds add a nutty taste and aroma.*

Aubergine 'fettucine'

Serves 4

- 400 g (14 oz) aubergines
- 4 g (1 tsp) salt
- 85 g (2 ¾ oz) onions
- 185 g (6 ½ oz) tomatoes
- Chilli pepper according to taste
- 2 cloves of garlic, peeled
- Leaves from 2 sprigs of thyme
- 15 g (½ oz) pistachios
- 3 tbsp olive oil
- 30 g (1 oz) sun-dried tomatoes
- Salt and pepper
- Parmesan cheese, grated

1. Trim the tops of the aubergines. Using a mandoline, cut off slices from the outer parts of the unpeeled aubergines, leaving the inner parts with the seeds. With a knife cut the slices into strips that resemble fettucine. Salt them and leave them in a cool place for at least 3 hours.
2. Cut the remaining inner parts of the aubergines into chunks.
3. Peel, trim, and cut the onions into segments.
4. Cut the tomatoes up into chunks. Chop the chilli peppers finely, mince the garlic, and strip the thyme leaves from the sprigs. Mix all of these together with the pistachio nuts and olive oil.
5. Spread out in a baking pan and bake at 200°C (390°F) for 15 minutes.
6. Add the sun-dried tomatoes and blend the mixture. Pour into a saucepan and season with salt and pepper. If the sauce is too thick add a little liquid.
7. Using your hands, squeeze as much moisture as possible out of the aubergine strips without breaking them. Cook them for 2–3 minutes in lightly salted water and then drain them.

Serving suggestions
Carefully mix the fettucine into the sauce and warm through for a couple of minutes. Serve immediately with grated Parmesan cheese sprinkled on top.

It is the slightly tough texture of the aubergines, similar to that of pasta, that holds this dish together. There is a great deal of umami from the tomatoes and the Parmesan cheese, and koku *sensation comes from the garlic.*

Serves 4

1 kg (4 ¼ c) simmered tomato sauce with herbs and vegetables (see Flavour Accelerators)
1 kg (2 lb 3 oz) zucchinis
1 kg (2 lb 3 oz) aubergines
Ca. 35 g (2 ⅓ tbsp) sea salt
500 g (1 lb 1 ½ oz) mushrooms, for example, chanterelles, porcini, and button mushrooms
Olive oil or butter
1 onion
1 lemon
2 cloves of garlic
3–4 sprigs of thyme
Salt and pepper
125 g (4 ½ oz) Parmesan cheese, grated

Green 'lasagna'
with aubergines, zucchinis, mushrooms, and tomato sauce

1. Prepare the simmered tomato, herb, and vegetable sauce and set aside.
2. Peel the zucchinis. Cut the aubergines and zucchinis into 1 cm (½ in) pieces lengthwise.
3. Score the aubergine slices in a diamond pattern half-way through and sprinkle with a little salt. Place them on a wire rack and allow the salt to draw out their moisture for about an hour.
4. Rinse the mushrooms and cut the bigger ones into large pieces.
5. Sauté the mushrooms in olive oil on high heat in a skillet, taking them out as they are done.
6. Pat the aubergine and zucchini slices dry, and sauté them as well.
7. Chop the onion finely, grate the peel from the lemon, and trim and mince the garlic cloves.
8. Toast the mushroom pieces a second time in a little butter, add the chopped onion, grated lemon peel, garlic, and thyme. Season with salt and pepper.
9. Cover the bottom of a baking dish with a thin layer of the sauce. Then place a layer of the zucchinis, aubergines, and mushrooms. Continue alternating the layers this way until everything is used up. Sprinkle the grated Parmesan cheese on top.
10. Cover the 'lasagna' with parchment paper and press it down gently with a smaller baking dish. Bake for at 165°C (330°F) for ca. 30 minutes.

Serving suggestions
Serve directly from the baking dish, possibly with sweet potatoes (see p. 250) on the side or with a green salad.

 This green 'lasagna' is a true umami feast with many different textures. The umami comes from the sauce and the Parmesan. Drawing out some of the moisture from the aubergines results in a slightly meat-like texture.

Serves 4

400 g (14 oz) fresh green asparagus

60 g (4 tbsp) capers

2 dl (⅘ c) neutral oil

100 g (3 ½ oz) *panko* or breadcrumbs

20 g (4 tsp) nutritional yeast flakes

A pinch of cayenne powder

Salt flakes

A small piece of fresh horseradish or a large radish

Green asparagus
with crisp capers, grated horseradish, and umami-crumble

1. Blanch the asparagus in salted water for 2 minutes. Remove them and plunge them into ice-cold water for a few minutes. Place them in a colander to drain.
2. Heat the oil in a skillet to a temperature of 165°C (330°F) and quickly fry the capers. Remove them with a slotted spoon and put them on absorbent paper towels.
3. On a dry skillet, toast the breadcrumbs, nutritional yeast, cayenne pepper, and some salt flakes until the mixture is light brown and gives off an intense aroma. Turn it all out onto a piece of parchment paper.
4. Peel the horseradish/radish.

Serving suggestions

Arrange the asparagus on a platter and sprinkle the crisp capers and toasted breadcrumb mixture on top. Grate a little horseradish or radish directly over the dish. Serve with country bread and a glass of good beer.

> *Even though this dish is deceptively simple, it benefits from the way the ingredients interact. The asparagus retain their own taste, which is like that of peas, while the acidity of the capers provides contrast to the umami from the nutritional yeast. The lightly cooked asparagus and the breadcrumbs contribute crispness in one way and the salt flakes add small crunchy touches. The strong spicy tastes of the cayenne powder and the horseradish give the dish the necessary kick that makes it interesting.*

Broccoli with *miso*-mayonnaise

Serves 4-6

1 head of broccoli
1 l (4 c) neutral oil
Salt flakes
Miso-mayonnaise
(see Flavour Accelerators)

1. Remove the leaves and coarsest bottom part of the stalk.
2. Cut the broccoli up into florets with the smaller stalks as well as pieces of the bigger ones still attached.
3. Put the oil in a pot and heat it to 170°C (340°F). Deep fry the florets until they are light brown. Work with a few florets at a time in order for the temperature of the oil to stay constant.
4. Remove the broccoli pieces, put them on absorbent paper towels, and sprinkle with the salt flakes.

Serving suggestions

Place the broccoli pieces in a bowl and serve with the *miso*-mayonnaise on the side. This dish can be used as an appetiser or as a side dish. The remaining stalk can be peeled or grated for use in salads or added to the bag of 'greens to go.'

The deep-fried broccoli pieces are exceptionally crunchy, and the miso-*mayonnaise adds umami and a creamy texture.*

Crisp cabbage
with asparagus and button mushroom fricassee

Serves 4

8 or more large outer leaves of a cone cabbage or a napa cabbage

Generously salted water

8–12 white asparagus

300 g (10 ½ oz) button mushrooms

1 l (4 c) lightly salted water

3 dl (1 ⅕ c) double or whipping cream

1 bunch of chives

Rustic croutons

1 slice day old bread

2 tbsp olive oil

1 clove of garlic

1. For the croutons: Rip the bread up into small, not too regular pieces. Crush the garlic with the back of a knife. Heat up a skillet with a little olive oil and the garlic and toast the bread pieces in it until they change colour and are a little burnt at the edges. Remove them from the skillet and place them on absorbent paper towels. Discard the garlic clove. Season with salt and pepper.
2. Separate the large leaves from the cabbage. Put the rest of the head aside for another use.
3. Bring a pot of generously salted water to a boil. Add the cabbage leaves and cook them until they are soft and well done.
4. Remove the leaves with a slotted spoon and let them dry completely on a cloth or paper towels. Set aside the cooking water.
5. Place the same number of oven-proof bowls as there are leaves upside down on a baking sheet or two (according to size) and cover them with the leaves. Bake them at 50°C (125°F) for ca. 60 minutes with the oven fan on (if there is one). Open the oven once and lift up the cabbage leaves so that they do not stick to the bowls.
6. Take the cabbage leaves out of the oven and leave them at room temperature until needed.
7. Peel the white asparagus and cut off the end pieces. Bring the peels and the end pieces to a boil in a pot with just enough of the cabbage water to cover. Turn off the heat and leave them in the water for ca. 10 minutes.
8. Remove the end pieces and the peels and discard them.
9. Cut the bottom one-third of the asparagus stalks into small pieces and return them to the pot.
10. Clean the button mushrooms. Cut one-third of them into quarters and put them in the pot as well.
11. Add lightly salted water so that there is ca. 1 l (4 c) liquid in the pot and allow the vegetables to simmer uncovered until the liquid is reduced by two-thirds.
12. Allow the cooked vegetables to cool a bit, leave them in the pot, and then use a hand blender to purée them.
13. Add the cream and heat until the sauce has thickened and is creamy.
14. Cut the asparagus into pieces ca. 3 cm (1 ¼ in) long and slice the remaining button mushrooms. Put them in the pot with the sauce and let them boil for ca. 3 minutes. If the sauce is too thin, remove the asparagus and button mushrooms with a slotted spoon, and allow the sauce to reduce until it has the desired consistency. Return the vegetables to the pot. This is now a vegetable fricassee.
15. Chop the chives finely.

Serving suggestions

Place the crisp cabbage leaves on plates to act as bowls and spoon the fricassee into them. Sprinkle the chives and croutons on top. Serve any remaining fricassee on the side.

This recipe can be seen as an interpretation of traditional, small savoury puff pastry tarts. Note that, even without any oil, the cabbage leaves become so crisp that they can be snapped in two, just like a puff pastry shell.

Cone cabbage with asparagus
and baked salted almonds, anchovy sauce, and oregano

Serves 4

- 12 white asparagus spears
- 12 green asparagus spears
- 1 cone cabbage or savoy cabbage
- Fresh oregano
- Salt flakes

Salted almonds
- 50 g (1 2/3 oz) almonds
- 2 dl (4/5 c) water
- 4 tsp salt
- Anchovy sauce (*bagna càuda*, see Flavour Accelerators)
- Bread

1. For the salted almonds: Dissolve the salt in the water, bring it to a boil and put the almonds in the water. Turn off the heat and leave the almonds in the water for about 1 hour.
2. Remove them from the water, spread them out on baking sheets, and bake in an oven at 165°C (330°F). Crush them after they have cooled.
3. Peel the white asparagus and trim off the tough bottom end. Check to determine whether the green asparagus also need to be trimmed.
4. Bring a pot of lightly salted water to a boil and cook the white asparagus for only about 3 minutes so that they remain crisp. Remove them with a slotted spoon and plunge them into ice-cold water.
5. Using the same water, cook the green asparagus for 1–2 minutes. As above, remove them with a slotted spoon and plunge them into ice-cold water.
6. Immerse the whole cabbage head for a few seconds in the boiling water and then plunge it as well into ice-cold water.
7. Remove all the vegetables from the ice baths and allow the water to drip off. Dry them in a cloth or with absorbent paper towels.

Serving suggestions

Cut the cabbage part way through lengthwise into several segments and place it on a large platter. Gently separate the leaves so that the cabbage is opened like a flower. Cut the asparagus stalks into medium sized pieces and distribute them among the leaves and around the cabbage. Crush the salted almonds and sprinkle them on top. Drizzle with the anchovy sauce and finish with the fresh oregano leaves and a few salt flakes. Serve extra anchovy sauce on the side together with crusty bread or small cooked potatoes.

The lightly cooked asparagus contribute a crisp mouthfeel, augmented by the crushed almonds. The green asparagus are rich in umami, further enhanced by the anchovy sauce, which is the very essence of umami. When asparagus are not in season, substitute other vegetables that can be served raw or only lightly cooked, such as black salsify, kohlrabi, radishes, or broccolini.

Dry-cured carrots *à la gravlax*

300 g (10 ½ oz) carrots, preferably large fairly thick ones

Marinade
2 ½ dl (1 c) water
25 g (5 tsp) brown sugar
1 stalk of dill
1 tsp fennel seeds
Pink peppercorn sauce (see Flavour Accelerators)
Fennel tops
Walnuts, preferably fresh

1. To make the marinade, boil together the water, brown sugar, dill, and fennel seeds for about 5 minutes. The amount of marinade required will depend on the size of the preserving jar and the volume of the carrots.
2. Peel and trim the carrots and cook them with the lid on until they are fork tender.
3. Place the carrots in a tall, slender jar or a wide-mouth canning jar. Pour the marinade over the carrots and store them in a cool place. They will easily keep for a week in the refrigerator.

Serving suggestions
Crush the walnuts. Cut the carrots into very thin slices lengthwise. Serve with fennel tops, the pink peppercorn sauce, and crushed walnuts.

> *The carrots are very tender and their natural sweetness is balanced by the spices. The walnuts add a little umami and some additional texture. The dish is meant to mimic both the appearance and taste of Nordic salmon* gravlax.

Carrots in zucchini sauce

Serves 4

800 g (1 lb 12 oz) different types and colours of carrots
10 g (2 tsp) nutritional yeast flakes
2 sprigs of thyme leaves, stems removed
Salt and pepper
50 g (1 ⅔ oz) butter

Zucchini sauce
500 g (1 lb 1 ½ oz) zucchinis
4 tbsp olive oil
3 cloves of garlic, peeled
2 anchovy fillets
Salt and pepper
½ bunch of parsley, chopped or *gremolata* (see Flavour Accelerators)

1. Scrape or peel the carrots and cut them up into randomly sized pieces.
2. Cut two large pieces of parchment paper to make a double layer. Place the carrots in a pile in the middle. Sprinkle the nutritional yeast, thyme leaves, salt and pepper on top, together with the butter. Fold up the sides and close them like a little bag. Use string if necessary to keep it closed.
3. Bake the carrots in the bag at 200°C (390°F) for ca. 12 minutes.
4. For the zucchini sauce: Cut the zucchinis up into large chunks and peel the garlic.
5. Pour the oil into a pot, add the garlic and let it sizzle a little. Add the zucchini pieces and anchovies and sauté until the zucchini is done.
6. Pulse with a hand blender and season with salt and pepper. Add a little water if the blended zucchini is too thick.

Serving suggestions
Divide the carrots among 4 deep bowls, making sure to distribute the melted butter mix in the bag as well. Pour the zucchini sauce on top and sprinkle the chopped parsley or *gremolata* on top.

> *The zucchini sauce is the most interesting part of this dish. Thanks to the anchovies it is super-charged with umami. The yeast flakes add more umami and the garlic creates a* koku *sensation. The zucchini sauce can also be used in other dishes, for example, with* pasta.

Dry-cured carrots *a la gravlax*.

Baked root vegetables
with tahini cream, basil, and mint

Serves 4

Dressing
1 chilli pepper
75 g (⅓ c) tahini
Worcestershire sauce, to taste
1 tbsp honey
200 g (⅞ c) crème fraîche
A little fine salt
20 stalks of basil
4–5 stalks of mint

Root vegetables
1 ½ kg (3 ½ lb) mixed root vegetables, for example, parsnips, carrots, celery, beets, sunchokes
1 sprig of thyme
½ dl (⅕ c) olive oil
2 organic lemons
Salt and pepper

Dressing
1. Cut open the chilli pepper, remove the seeds and the membranes, and chop it very finely.
2. Combine the tahini, Worcestershire sauce, and honey. Mix in some or all of the chilli pepper according to how hot the sauce should be.
3. Mix in the crème fraîche and season with salt.
4. Chop half of the basil and mint leaves and mix them in. Set aside the remaining herbs for serving the dish.

Root vegetables
1. Pre-heat the oven to 200°C (390°F).
2. Trim and peel the vegetables and cut them into pieces that are suitable for their shape. Or, in the case of small ones, leave them whole.
3. Remove the thyme leaves from the stalk. Toss the vegetables with the olive oil and mix in the thyme leaves. Grate the peel from the lemons and mix in together with salt and pepper.
4. Spread the vegetables out in a baking dish and bake for ca. 25–30 minutes.

Serving suggestions
Transfer the warm vegetables to a platter or serve directly from the baking dish. Drizzle a little of the dressing on top. Garnish with the remaining basil and mint leaves.

This is a dish that can really boost one's vegetable intake. Baking releases more of the sweetness in the root vegetables and their various colours and shapes are a feast for the eyes. The dressing has sweetness from the honey, a little umami and spiciness from the Worcestershire sauce, and pleasing aromas from the sesame-based tahini and the herbs.

Baked sunchokes
with almond milk mashed potatoes

Serves 4

800 g (1 lb 12 oz) baking potatoes
600 g (1 lb 5 oz) sunchokes, after peeling
6 sprigs of thyme
Olive oil
Salt and pepper
1 organic lemon
1 bunch of parsley
Salt flakes

Almond milk
6 dl (2 2/5 c) water
150 g (5 1/3 oz) almonds
or ca. 6 dl (2 2/5 c) commercial almond milk

1. If not using commercial almond milk: Bring the water to a boil in a saucepan and leave the almonds in it for 2 minutes. Remove the almonds and skin them. Blend them thoroughly with the cooking water.
2. Peel the potatoes, cut them into large pieces, and cook them in lightly salted water until they are fork tender. Drain them, reserving the potato cooking water for later use.
3. Pre-heat the oven to 200°C (390°F).
4. Clean the sunchokes. Peel them if the peel is tough or if they are very difficult to clean properly.
5. Remove the leaves from the thyme sprigs. Toss the sunchokes with olive oil, the thyme leaves and stalks, salt, and pepper. Spread them out in a baking dish and bake for ca. 20 minutes, depending on their size and how firm they should be. Discard the thyme stalks.
6. Heat the almond milk while stirring constantly so that it does not separate.
7. Mash the potatoes and mix in the almond milk. Season with salt and pepper. If necessary, add a little of the potato cooking water so that the potatoes have a soft, velvety, and creamy texture.
8. Rinse and chop the parsley.

Serving suggestions
Arrange the potato mash in the middle of four plates and arrange the warm sunchokes on top. Garnish with the chopped parsley, grate some lemon peel over the sunchokes, and finish with a few salt flakes. Serve immediately.

This is a light, simple vegan dish that features the rustic-looking sunchokes on the smooth, white background of the mashed potatoes. The overall effect is creamy with some crispness from the baked peel and edges of the baked sunchokes. The almond milk enhances the natural nutty taste of the sunchokes.

1 large *daikon*

Dashi
(see Flavour Accelerators)

Soy sauce

Simmered *daikon*

1. Peel the *daikon* and cut into rounds about 3–4 cm (1 ¼–1 ½ in) thick. Use the thickest end of the root.
2. Place the rounds in a pot with enough *dashi* to cover and simmer them with the lid on for ca. 20 minutes until they have become a little soft. Bouillon can be substituted for the *dashi*.
3. Drain the *daikon* rounds. Arrange them on a plate and drop a little soy sauce in the centre of each. It will seep out decoratively along the fibres of the *daikon*.

Serving suggestions

The *daikon* rounds can be eaten cold as a snack or as a side dish. For example, they can be served cold with cooked kalettes and grated lemon rind or with freshly peeled, cooked shrimp.

> *This recipe will come as a big surprise for many and shines a completely new light on the* daikon *as a delectable vegetable. This otherwise hard, crisp Chinese radish becomes soft and succulent, with a great deal of umami from the* dashi *and soy sauce.*

Red beets and kidney beans
with dried duck breast and cherries

Serves 4

200 g (7 oz) kidney beans
500 g (1 lb 1 ½ oz) red beets
½ dl (⅕ c) red wine vinegar
4 medium sized red onions
A little olive oil
40 g (1 ⅓ oz) dried cherries
1 dl (⅖ c) cherry wine or sweet red wine
Salt and pepper

Dried duck breast
2 duck breasts, ca. 300–400 g (⅔–1 lb)
20 g (⅔ oz) dried cherries
20 g (4 tsp) sea salt
2 g (½ tsp) freshly ground black pepper

Duck breast
1. If the fat layer on the duck breast is very thick, cut some of it off.
2. Chop the dried cherries and mash them in a mortar with the salt and pepper to make something like 'cherry salt.'
3. Place the duck breasts with the cherry salt in a sealable bag and place it in the refrigerator. Shake it up a bit every day for 3 days.
4. Remove the duck breasts and wipe them off thoroughly. Place them in a clean cheesecloth and hang them up in a cool place or in the refrigerator for 5–10 days, depending on how firm one wants the meat to be.
5. This recipe makes more than is needed for this dish, as it is not worthwhile to make it in a smaller quantity. The dried duck breast can be used in other dishes and will keep under refrigeration for a week or longer in the freezer.

Red beets and kidney beans
1. Soak the beans the day before.
2. Cook the beans in lightly salted water according to the directions on the packet. Drain off the water and reserve a little of it. Leave the beans in a colander to let the last of the water drain away.
3. Cook the beets for 30–45 minutes in lightly salted water until they are done, but still firm. Drain the cooking water and place the pot under cold running water until the peels are easy to slip off, but the beets are still warm.
4. Remove the peels and cut the beets up into bite-sized, irregular pieces. Season with salt and place them in the red wine vinegar.
5. Trim and peel the onions, cut them in half lengthwise, and drizzle with a little olive oil. Bake them for ca. 20 minutes in the oven at 165°C (330°F).
6. When they are done, separate the onions into cup-shaped layers. Season with a little salt.
7. Place the cooked beans in a saucepan with the beets, vinegar, cherries, and cherry wine. Mix them all together and season with salt and pepper.

Serving suggestions
Warm the beans/beets mixture and taste it to ensure that the salt-acid balance is correct. Carefully stir in the onions to make a sort of ragout. Distribute it on plates or arrange on a platter. Cut the duck breast into paper-thin slices and arrange them around the beets or serve them on the side.

> *In this recipe, a small amount of meat transforms the vegetable dish by adding umami and texture contrast that result in a fantastic sensation of mouthfulness. The beets combined with the red wine vinegar and the cherry wine complement the nutritious kidney beans that are otherwise a little dry and bland. If the dish is to be vegan, mushrooms such as king oysters or Portobello can replace the duck breasts.*

Serves 4

4 small, spring onions with roots and tops

Extra virgin olive oil

Salt and freshly ground pepper

100 g (3 ½ oz) small button mushrooms

2–3 fresh chive stems, chopped finely

1 piece day old sourdough bread

Sea salt flakes

16 small anchovy fillets, preferably in olive oil

1 small piece of Parmesan cheese

1 small piece of frozen *pata negra (jamón ibérico)* or other air-dried ham

Balsamic vinegar

Grilled onions
with sourdough bread, anchovies, mushrooms, and *pata negra* ham

1. Clean and trim the onions. Toss them with a little olive oil, salt, and pepper.
2. Heat a grill or a grill pan to a high temperature. Quickly grill the onions.
3. Cut the mushrooms up into thin slices and toss them with the chopped chives and a few drops of olive oil.
4. Break the bread up into smaller chunks and sauté them in a little olive oil on a skillet until they are golden.
5. Sprinkle sea salt over them and place them on absorbent paper towels.

Serving suggestions

Arrange the bread chunks, anchovies, grilled onions, and mushrooms on plates. Grate the frozen ham on top and sprinkle with grated Parmesan cheese. Finish with a few drops of balsamic vinegar and olive oil.

> *The rich taste of this dish comes almost entirely from the umami synergy generated by the anchovies, Parmesan cheese, mushrooms, balsamic vinegar, and the aged, air-dried ham. Koku sensation comes from the onions. The ham can be omitted if the dish is to be vegan-friendly.*

Serves 4

8–12 leeks, depending on size
100 g (3 ½ oz) feta, crumbled
50 g (3–4 tbsp) capers
Potato cooking water or *dashi* (see Flavour Accelerators)

Dressing

200 g (⅞ c) Greek yogurt, 3% or 10%
2 red or green bell peppers
1 red chilli pepper
1 clove of garlic
1 organic lemon
50 g (1 ⅔ oz) feta, crumbled
A handful of fennel tops or possibly basil leaves
1 small bunch of mint
Salt and pepper

Succulent leeks
with feta, bell pepper yogurt, mint, and capers

Dressing

1. Drain the yogurt in a paper coffee filter so that it becomes firmer. It will take 2–3 hours for the 3% yogurt and about 1 hour for the 10% yogurt.
2. Place the whole bell peppers on a baking pan lined with parchment paper. Bake in a preheated oven at 220°C (430°F) for 15–20 minutes. Transfer them to a closed plastic bag and leave them for 20 minutes. Peel and remove the seeds.
3. Chop up the peppers and the chilli pepper and mix with the yogurt. Put the garlic through a garlic press and grate the lemon peel. Stir the garlic and lemon peel, along with the crumbled feta, into the yogurt. Chop up the fennel tops and mint leaves and mix in half of them. Season with salt and pepper.

Leeks

1. Carefully cut off the roots of the leeks, keeping as much of the white end as possible. Trim off the very top ends of the green part.
2. Rinse the leeks thoroughly to wash away any dirt caught between the leaves by making lengthwise partial incisions in the green part that may contain dirt.
3. Cook the leeks in lightly salted water, potato cooking water, or *dashi*. They should be soft but still a little firm. Test this with the tip of a paring knife to avoid overcooking them.
4. Remove the leeks and allow them to drain on absorbent paper towels.

Serving suggestions

The leeks can be served hot, warm, or cold. Arrange them whole or cut on a slant into 3–4 pieces on a platter. Sprinkle first the crumbled feta and capers on top, followed by a lot of chopped fennel and mint. Serve the dressing on the side.

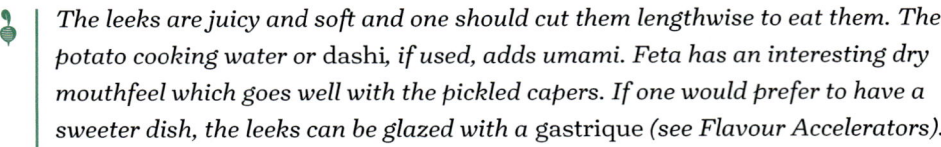

The leeks are juicy and soft and one should cut them lengthwise to eat them. The potato cooking water or dashi, *if used, adds umami. Feta has an interesting dry mouthfeel which goes well with the pickled capers. If one would prefer to have a sweeter dish, the leeks can be glazed with a gastrique (see Flavour Accelerators).*

Serves 4

1 kg (2 lb 3 oz) onions
A little butter
1 kg (2 lb 3 oz) mealy, peeled potatoes
2 ½ dl (1 c) milk
40 g (2 ⅔ tbsp) butter, at room temperature
300 g (10 ½ oz) kale or crisp savoy cabbage
Ca. 200 g (7 oz) tart apples
A little lemon juice
Salt and pepper

Chilli dates
150 g (5 ⅓ oz) large, soft, pitted dates
1 small chilli pepper
½ dl (⅕ c) soy sauce
1 tsp dark *miso* paste, either red or *mugi* (see Flavour Accelerators)

Potato purée
with kale, apples, onions, and chilli dates

1. Chop the chilli very finely, dissolve the *miso* paste in the soy sauce, put the dates in a zip-lock bag, and mix everything together.
2. Leave on the counter for a couple of hours or overnight in the refrigerator. If refrigerated, bring the dates to room temperature before serving.

Potato purée
1. Peel and trim the onions and cut them up into slices. Place in a pot at low heat with the butter and let them simmer for 60 minutes until they are soft and have turned golden. Season with salt.
2. Cut the potatoes up into large chunks and cook them in lightly salted water until they are done.
3. Drain and return to the heat briefly to steam off the remaining moisture.
4. Bring the milk to a boil, add to the potatoes, and mash them with the butter.
5. Season the purée with salt and pepper and keep warm.
6. Rinse the kale and blanch it quickly in boiling water.
7. Remove the kale with a slotted spoon and plunge it into cold water with ice cubes. Then spin it or place it on absorbent paper towels so that is as dry as possible.
8. Wash the apples, cut them in half and core them. Chop them up into small cubes and sprinkle them with lemon juice to prevent them from turning brown.
9. Season the kale with a little salt and mix in the apple pieces.

Serving suggestions
Divide the kale-apple mixture among four plates or arrange on a platter. Cover with the potato purée and soft onions. Distribute the chilli dates placing them on top or at the side, with a little of the juice in which they were marinated.

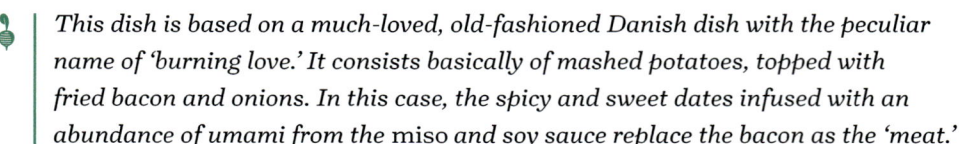

This dish is based on a much-loved, old-fashioned Danish dish with the peculiar name of 'burning love.' It consists basically of mashed potatoes, topped with fried bacon and onions. In this case, the spicy and sweet dates infused with an abundance of umami from the miso *and soy sauce replace the bacon as the 'meat.'*

Serves 4

800 g (1 lb 12 oz) different varieties of potatoes

10 g (⅓ oz) *konbu*

3 dl (1 ⅕ c) potato cooking water

100 g (3 ½ oz) fresh goat cheese

2–3 different types of cress, for example, water cress, garden cress, pepper grass

Potatoes with a difference
with fresh goat cheese and a lot of cress

1. Wash the potatoes thoroughly and peel them or scrape off the skin on new ones. Cook them in lightly salted water with the *konbu* until they are tender.
2. Drain off the cooking water, reserving 3 dl (1 ¼ c) of it. Discard the seaweed or keep it for a salad.
3. Pour the potato cooking water back into the pot together with the potatoes. Add the goat cheese, letting it dissolve to make a creamy sauce. Bring to a boil and let the sauce reduce until it has thickened a little.
4. Rinse the cress and chop the larger pieces coarsely, if necessary.

Serving suggestions

Cut the larger potatoes in half. Divide the potatoes and sauce among four deep bowls, sprinkle with the cress, and serve immediately.

> *This dish celebrates potatoes—tasty vegetables that, together with the water in which they are cooked, are full of umami, which is further enriched by the umami from the* konbu. *The cress has a bitter and sharp taste. The creamy consistency of the goat cheese pulls it all together.*

Potatoes with fresh cheese and raw mushrooms

Serves 4

8–12 medium sized baking potatoes, unpeeled

Coarse salt for baking

Salt and freshly ground pepper

15–20 g (3–4 tsp) nutritional yeast flakes

1–2 dl (²/₅–⁴/₅ c) fresh soft cheese, for example, ricotta or creamy goat cheese

1 handful of flat-leaf parsley

A little truffle, a large porcini mushroom, or two button mushrooms

Neutral oil for frying

Fine salt and salt flakes

1. Thoroughly scrub the potatoes. Toss them with coarse salt and place them in a baking dish. Bake at 160°C (320°F) until they are done.
2. Cut the tops off the potatoes and spoon out the centres, leaving a little layer of potato next to the skin, which is to be fried later. Place the cooked potato in a bowl.
3. Season the potato with salt and pepper. Gently stir in the yeast flakes and half of the fresh cheese. Do not overmix as the potatoes should retain their shape.
4. Chop the parsley and clean the mushrooms.
5. Heat oil to 165°C (330°F) in a pot and fry the potato skins until they are golden and crisp. Place them bottom up on absorbent paper towels and sprinkle a little fine salt over them. The skins are left at room temperature until they are served; they would become limp in the refrigerator.

Serving suggestions

Stuff each potato skin with a spoonful of the cooked potato and a little extra fresh cheese. Using a mini-mandoline, shave the truffle and mushrooms over them. Sprinkle the chopped parsley and salt flakes on top.

This dish can be thought of as a warm potato salad, but it is actually a tribute to the humble potato skin, a part of the tuber that is often discarded. Here it is transformed to end up with a rich umami taste and a crisp texture. Extra umami comes from the yeast flakes. The truffle shavings add even more umami, as well as releasing enticing aromas. The shaved mushrooms are a little crunchy, while the creamy fresh cheese holds it all together.

Pears, beans, potatoes, and smoked pork belly

Serves 4

600 g (1 lb 5 oz) potatoes
400 g (14 oz) green beans
200 g (7 oz) onions
1 ½ l (6 c) vegetable bouillon, potato cooking water, or water (see Flavour Accelerators)
5–6 sprigs of thyme
2 g (½ tsp) crushed black peppercorns
200 g (7 oz) smoked pork belly or bacon
500 g (18 oz) pears
1 bunch of parsley

1. Peel the potatoes, trim the green beans, and cut the onions up into small sections.
2. Bring the liquid to a boil, add the thyme sprigs and crushed peppercorns.
3. Slice the smoked pork belly and cook the slices in the liquid until they are tender, about 20–25 minutes.
4. Remove the meat and set it aside on a plate.
5. Cook the potatoes, beans, onions, and possibly some additional thyme sprigs in the bouillon on low heat for about 10 minutes.
6. Slice the pears in half and remove the core. Cut each pear into 6 or 8 lengthwise segments.
7. Place the pears in the pot with the vegetables and continue to cook for another 20 minutes.
8. Rinse and chop the parsley.

Serving suggestions
Place the pork belly back in the pot just before serving and warm it through. Serve the dish in the pot or on a rustic deep platter, with the parsley sprinkled on top. The pears can also be left whole and simmered with the vegetables from the start.

Choose pears that are firm and somewhat grainy, but juicy and sweet. Bosc and Anjou are possible candidates. The pear slices can also first be grilled on a wood-fired grill to increase the smoky character of the dish. The pork belly adds smoky aromas, salty, as well as umami taste substances.

Serves 4

50 g (1 ⅔ oz) dried, pitted olives, preferably Kalamata

Ca. 1 kg (2 ¼ lb) large sweet potatoes

3 tbsp pumpkin seed or olive oil

1 tsp dried chilli flakes

Sea salt

50 g (1 ⅔ oz) pumpkin seeds with soy sauce (see Flavour Accelerators)

Goat cheese cream

2 dl (⅘ c) skyr or crème fraîche

50 g (3 ½ tbsp) creamy goat cheese

Pinch of cayenne pepper

Sweet potatoes
with chilli, goat cheese cream, and pumpkin seeds

1. Cut the olives in half, place them on parchment paper on a baking sheet, and bake them at 50°C (125°F) for about 2 hours until they have dried out.
2. Rinse the sweet potatoes and cut each of them, unpeeled, into 4-6 segments lengthwise, depending on their size. Rub them with oil, the chilli flakes, and sea salt. Place them on a baking sheet lined with parchment paper.
3. Bake the sweet potatoes at 210°C (410°F) for 15–20 minutes until they are fork tender.
4. Mix together the skyr and the goat cheese. Season with cayenne pepper and salt.
5. Chop the dried olives.
6. Pour the goat cheese cream into a bowl and sprinkle the olive bits on top.

Serving suggestions

Cut the sweet potato segments into smaller pieces. Crush the pumpkin seeds and sprinkle them on the warm sweet potatoes. Serve with the goat cheese cream on the side. This dish can also be paired with green 'lasagna' (see p. 218).

The sweet potatoes are the focal point of this dish. The soy-roasted pumpkin seeds contribute umami and a crunchy mouthfeel. The chilli flakes spice things up but are balanced by the mild taste of the goat cheese cream. This dish is versatile and goes well with many other vegetables.

Couscous

with *za'atar*, Brussels sprouts, and roasted cauliflower

Serves 4

3 dl (1 ⅓ c) water
1 ½ dl (⅔ c) couscous
1 tsp salt
A little olive oil
A little sea salt
1 clove of garlic, crushed
400 g (14 oz) cauliflower
400 g (14 oz) Brussels sprouts
40 g (2 ⅔ tbsp) butter
2 tsp sesame oil
4 tbsp olive oil
2–3 tsp *za'atar* (see Flavour Accelerators)
Parsley and cilantro

1. Bring 3 dl (1 ⅕ c) water to a boil, add the salt, and stir in the couscous. Cover with the lid of the pot and let it stand for 15 minutes.
2. In a bowl mix together a little olive oil, salt, and the crushed garlic. Break the cauliflower apart into florets and toss with the oil mixture.
3. Bake the cauliflower on a sheet pan for 6–10 minutes in the oven at 180°C (355°F). The florets might char a little, but should remain firm. Remove from the oven and keep them covered.
4. Rinse and trim the Brussels sprouts. Cut them in half lengthwise, and put them in a pot with 1 dl (⅖ c) water.
5. Bring the pot to a boil, add 40 g (2 ⅔ tbsp) butter, cover the pot and let the sprouts cook on high heat for 5 minutes.
6. Mix the warm couscous with the sesame oil and mix it into the pot with the Brussels sprouts.
7. While leaving the pot on the stove, stir the couscous mixture, mashing the sprouts a bit with the cooking spoon, so that everything is well mixed together. Season to taste with salt as needed.
8. Chop the parsley and cilantro coarsely.

Serving suggestions

Spoon the couscous/Brussels sprout mixture into pasta bowls together with the cauliflower florets. Drizzle with 4 tablespoons of olive oil and sprinkle the *za'atar*, parsley, and cilantro on top.

 The combination of garlic, sesame oil, and the special za'atar spice mixture transforms this simple vegetable dish into a complete main course. It is also easy to add more za'atar and salt to taste at the table.

Serves 4

Quinoa salad with dulse

10 g (⅓ oz) pieces of dried dulse

Soy sauce

600 ml (⅗ c) water

250 g (8 ⅔ oz) quinoa

1. Soak the dulse in water with a little soy sauce.
2. Rinse the quinoa thoroughly in cold water. Toast it in a dry pot at low heat until it turns light brown. This removes some of the bitter substances in the shells.
3. Cook the quinoa in water on low heat for ca. 20–25 minutes. Pay attention so that all the water does not cook away.
4. Drain the quinoa, cool it, and mix with the dulse.

> *Dulse is one of the types of seaweed that can contribute umami. Here it is used to enhance the taste of a simple quinoa salad. Prepared this way, the dulse can also serve to add an extra touch to mashed potatoes or an omelette.*

Quinoa, cauliflower, and tomatoes
with a tangy citrus dressing and crispy cod

Serves 4

Dressing
2 spring onions
1 dl (²/₅ c) freshly squeezed orange juice
½ dl (⅕ c) freshly squeezed lime juice
¼ dl (1 ⅔ tbsp) freshly squeezed lemon juice
1 tbsp Worcestershire sauce
3 tbsp tomato ketchup
½ tsp honey
¼ tsp ground pepper

Quinoa and vegetables
250 g (1 c + 1 ½ tbsp) quinoa
1 small cauliflower, ca. 700 g (1 ½ lb)
2 tsp olive oil
Salt
125 g (4 ½ oz) almonds
A little olive oil
½ tsp fine table salt
400 g (14 oz) tomatoes
1 cucumber, ca. 300 g (10 ½ oz)
1 bunch of parsley
A little fine table salt
Pepper

Cod
2 tbsp sesame seeds
½ tbsp sea salt
2 tsp dried oregano
1 tsp crushed coriander seeds
Pinch of cayenne pepper
4 pieces of cod or other white fish about 100 g (3 ½ oz) each
A little olive oil
A little butter

A small amount of fish elevates this dish of quinoa and cauliflower to another level. The fish, tomatoes, Worcestershire sauce, and ketchup contribute umami, while the crushed almonds are a source of crunch.

Dressing
1. Slice the spring onions into very thin rounds.
2. Mix together the other ingredients and add the spring onions.

Quinoa and vegetables
1. Pre-heat the oven to 230°C (445°F).
2. Cook the quinoa in lightly salted water according to directions on the package.
3. Drain, allow the quinoa to cool, and transfer to a large bowl.
4. Cut the cauliflower up into very small florets and toss them with the olive oil and salt. Place them on a baking tray lined with parchment paper.
5. Toss the almonds with a little olive oil and salt and place them on another baking tray.
6. Bake the cauliflower florets until they have charred a little but are still a bit firm and the almonds until they are golden brown. Note that as the oven is very hot, this may take only a very short time and the almonds, especially, should not be allowed to burn and may need to be taken out first. Remove from the oven and allow them to cool down.
7. Cut the tomatoes and cucumber, skin and seeds included, into small cubes. Add to the quinoa together with the cauliflower.
8. Chop the parsley very finely, add to the quinoa as well, and gently mix everything together.
9. Season to taste with salt, pepper, and a generous amount of the citrus dressing.
10. Crush 10–15 almonds coarsely. Set aside any leftover almonds for later use in a jar with a tight-fitting lid.

Cod
1. Mix together the sesame seeds, sea salt, oregano, coriander seeds, and cayenne pepper in a small bowl.
2. Dry off the fish pieces with absorbent paper towels.
3. Coat the fish with the seasoning mixture, rubbing it on thoroughly to make it stick.
4. Heat a little olive oil and butter in a skillet at medium heat and sauté the fish until it is done and the crust is crisp and golden brown.

Serving suggestions
Arrange the salad on a platter or in a wide, shallow bowl. Place the fish pieces on top and sprinkle with the crushed almonds.

Polenta fritters with blue cheese dip

Serves 4

2 ½ dl (1 c) instant polenta
2 tsp dried or freshly chopped oregano
Salt and pepper
60 g (2 oz) Parmesan cheese
150 g (5 ⅓ oz) mozzarella or other semi-soft cheese such as Taleggio
2 eggs
Breadcrumbs, preferably *panko*
Neutral oil for frying
Salt and a little cayenne pepper
1 dl (⅖ c) milk
150 g (5 ⅓ oz) blue cheese, for example, Gorgonzola or Danish Blue

1. Prepare the instant polenta according to the directions on the package, adding the oregano right away and seasoning with salt and pepper at the end.
2. Grate the Parmesan finely and the mozzarella coarsely and stir both cheeses into the polenta.
3. Line a baking pan of suitable size, for example 20 × 30 cm (8 × 12 in), with plastic wrap. Spread the polenta evenly out in the pan, going right to the edges. It should be about 1 ½–2 cm (½–¾ in) thick.
4. Cover the surface of the polenta with another layer of plastic wrap and place the pan in the refrigerator for at least 3 hours.
5. Take the pan from the refrigerator, remove the top layer of plastic wrap and turn the polenta out onto a cutting board. Cut it into more or less even strips, ca. 1 ½ × 8 cm (⅔ × 3 ¼ in) wide.
6. Beat the eggs in one bowl and pour the breadcrumbs into another.
7. Dip the strips first in the egg and then in the breadcrumbs so that they are evenly covered.
8. Heat the oil in a deep pot to 165°C (330°F) and deep-fry the polenta strips a few at a time until they are golden and crisp. Place them on absorbent paper towels and sprinkle with fine salt and a little cayenne pepper.
9. Warm up the milk and blue cheese until the cheese has melted. Stir to mix it into the milk. Pour into a bowl and set aside.

Serving suggestions
Serve the warm polenta fritters with the blue cheese dip on the side. The fritters can also be sprinkled with salted mushroom powder (see Flavour Accelerators).

The fritters can be used as a side with a number of different vegetable dishes thanks to the Parmesan and blue cheese that supercharge their umami content. They can also be served as an appetiser.

Baked squash with mushroom risotto and crumble

Serves 4

1 small squash, for example, Hokkaido or butternut

Crumble, e.g., butter-roasted breadcrumbs

Risotto

40 g (1 ⅓ oz) shallots

3 sage leaves, chopped finely

30 g (2 tbsp) grated, fresh Parmesan cheese, plus extra for gratinéing

½ l (2 c) basic vegetable stock (see Flavour Accelerators)

2 tbsp olive oil

125 g (½ c) Arborio or other risotto rice

200 g (7 oz) mushrooms, for example, porcini, chanterelles, and field mushrooms

1 clove garlic

A little butter for frying

Salt and pepper

Risotto

1. Peel the shallots and chop them finely, chop the sage leaves, and grate the Parmesan cheese. Warm the vegetable stock to the boiling point.
2. Pour the olive oil into a saucepan, add the chopped onions and allow them to simmer until they are soft, stir in the rice and allow it to simmer with the onions for a little while.
3. Ladle in the warm stock, a little at a time, stirring constantly at first, and after that frequently. (Consult the rice package for details about the cooking time and amount of liquid to be added. These vary considerably from one brand to another.)
4. Rinse the mushrooms and cut them into smaller pieces.
5. Crush the garlic a little. Sauté it and the mushrooms in butter in a skillet and season well with salt and pepper. When the mushrooms are done, discard the garlic.
6. Stir the Parmesan cheese and chopped sage into the risotto. Add in the mushrooms and salt and pepper according to taste. Set the risotto aside to allow the flavours to marry.

Squash

1. Cut away the outer side on opposite sides of the squash, to give access to the seeds inside. Depending on the shape of the squash cut it into 4 thick uniform slices, either lengthwise or crosswise. Scoop out the seeds from the slices, leaving a hole in the middle. Place the slices on a baking sheet covered with parchment paper and season with salt and pepper.
2. Bake the squash slices at 185°C (365°F) for 8 minutes. They should be done, but still firm.
3. Just before serving, fill the holes in the squash with warm risotto. Sprinkle with a little grated Parmesan cheese and bake once again at 185°C (365°F) for 6–8 minutes.

Serving suggestions

Arrange the squash slices on plates and sprinkle crumble on top. Serve with a seasonal salad, for example, with kale, apples, cranberries, and fresh hazelnuts.

When paired with a kale salad, this dish can be used as a main course. The crumble adds a toasted, nut-like aroma and taste, while the risotto is rich in umami from the vegetable stock, mushrooms, and Parmesan. The garlic contributes koku *sensation.*

Pasta *tarako*

Serves 4

100 g (3 ½ oz) roe, fresh or preserved for example, from cod, salmon, or lumpfish

250 g (9 oz) spaghetti

A little cooking water

2 tbsp soy sauce

2 dl (⅘ c) cream

2 tbsp melted butter

2–3 tbsp lemon juice

1 spring onion or a few green *shiso* leaves

This delicately flavoured dish, traditionally made with *tarako* (pollock roe) is a Japanese specialty from the Fukuoka prefecture.

1. If using fresh roe, squeeze or scrape it carefully out of the membrane.
2. In a bowl mix together the roe, soy sauce, cream, melted butter, and lemon juice.
3. Cook the spaghetti in lightly salted water according to the instructions on the package, minus 1 minute. Drain the water, but reserve a little.
4. Chop up the spring onion or *shiso* leaves finely.
5. Transfer the cooked pasta to an appropriately sized skillet, together with a few spoonfuls of the cooking water and warm it up. Stir in the roe mixture until it coats the spaghetti and the dish looks creamy.
6. Sprinkle the spring onion or *shiso* on top and serve immediately.

> *In this dish the neutral-tasting pasta is the perfect foil for the essence of the tastes of the other ingredients and contributes a pleasing al dente texture.*

Mushroom pâté

750 g (1 lb 9 oz) fresh porcini, king oyster, or button mushrooms

25 g (1 oz) peeled garlic cloves

1 dl (2/5 c) olive oil

10 g (1/3 oz) salted anchovy fillet

Salt and pepper

1. Rinse, trim, and slice the mushrooms.
2. Crush with garlic cloves with the heel of the hand, place them in a cold saucepan with the olive oil, and sauté them on low heat until they are golden.
3. Add the anchovy fillets and let them simmer a little.
4. Stir in the mushroom slices and simmer until all the liquid has evaporated, stirring occasionally.
5. Pour the mushroom mixture into a blender and purée it. Add more oil as needed to obtain a softer consistency.
6. Season with salt and pepper. Transfer the pâté to sterilised jars and store in the refrigerator. If the pâté is to be kept for more than two weeks, it can be frozen in the jar. Or it can be pasteurised in the jar in a preheated oven at 75°C (165°F) for 25 minutes.

Serving suggestions

Use the mushroom pâté as a tapenade on toast, as a taste additive in a soup, in a pasta dish, or tossed with steamed vegetables.

The pâté is creamy and has an intense umami taste from the mushrooms and anchovy fillet. The garlic adds piquancy and brings out a sensation of koku.

Yellow split pea hummus with thyme

Serves 4

Ca. 450 g (1 lb) dried yellow split peas (or other legumes) to yield ca. 900 g (2 lb) when soaked
4 cloves of garlic
5–6 sprigs of thyme
2 bay leaves
80 g (2 ⅔ oz) Parmesan cheese rinds (ideal, but can be left out)
Pinch cayenne pepper
½ dl (⅕ c) olive oil
¼ dl (5 tsp) mushroom oil
Cooking water from the peas
Salt and pepper

1. Soak the split peas for 8 hours.
2. Place the peas in a saucepan and cover with water. Peel the garlic cloves, crush them with the heel of your hand or the flat side of a knife blade. Strip the thyme leaves from the stems. Add the garlic, thyme leaves and stems, bay leaves, and Parmesan rinds to the pot.
3. Cook the peas for 30–40 minutes. Drain the peas, reserving the cooking water. Discard the Parmesan cheese crust, bay leaves, and thyme stems.
4. Purée the peas with a little of the cooking water and cayenne pepper. Add oil and cooking water alternately, stopping when the consistency is creamy. Season with salt and pepper.

Serving suggestions

The hummus can be used as a condiment or sauce with many different vegetable dishes, as filling in a sandwich or pita pockets, or with toast or crispbread. Mushroom powder salt (see Flavour Accelerators) can be sprinkled on top.

> *Yellow split pea hummus is creamy and sweet with a well-rounded herb taste and* koku *sensation from the garlic. The Parmesan cheese crust contributes a great deal of umami.*

Lentils, romanesco, and pomelo

Serves 4

200 g (7 oz) lentils, e.g., red lentils

1 head of romanesco or possibly cauliflower or broccoli

100 g (3 ½ oz) cashew nuts

2–3 organic limes

Herbs, for example, parsley, wild garlic, or tarragon

A little crème fraîche or aspic (optional)

Salt

Dressing

1 pomelo

100 g (3 ½ oz) mild white onions, for example spring onions

50 g (ca. 2 oz) soft dates

1 dl (²⁄₅ c) orange juice

Extra lime juice to fine-tune the seasonings

A few chilli flakes

2 tbsp rice vinegar

½ dl (⅕ c) *ponzu* (or soy sauce with lemon juice) (see Flavour Accelerators)

Dressing

1. Peel the pomelo and separate the segments from the membranes. Working over a bowl to catch the juice for the dressing, cut up the flesh into smaller bite-sized pieces. Put aside.
2. Peel, trim, and chop the white onions, pit the dates, and mix them with the remaining ingredients and the pomelo juice. Blend to a thick consistency.

Lentils

1. Cook the lentils according to directions on the package and drain.
2. Cut the romanesco up into small florets.
3. Toast the cashew nuts on a dry skillet until they turn light brown. Chop them coarsely.
4. Grate the rind from the lime over the lentils. Season with their juice and salt.
5. Chop the parsley and mix it in.
6. Stir in the dressing. Adjust the seasonings with more acidity, sweetness, chilli, and salt. The dish is intended to be juicy and have a bit of a kick.

To serve

Arrange the lentils on a platter or in individual pasta bowls. Distribute the romanesco florets and pomelo pieces. Sprinkle the chopped cashew nuts on top and, optionally, drizzle with a little crème fraîche. The dish can be eaten as an entrée or as a side dish.

The dressing in this dish is very versatile and can be used in a number of other settings. The ponzu *contributes acidity from the* yuzu*, which is enhanced by the aromatic rice vinegar, while the soy sauce adds umami. The dates add sweetness and help to thicken the dressing. Pomelo is normally a somewhat dry fruit, but here it becomes juicy from absorbing the liquid in the dressing.*

White beans with seaweeds
three ways—as a salad, sautéed, puréed

Serves 4

Ca. 300 g (2/3 lb) dried white beans, when soaked, 600 g (1 1/3 lb)

15 g (1/2 oz) *konbu* seaweed

1 tsp salt

As a salad

200 g (7 oz) cooked whole beans

1 dl (2/5 c) sherry vinegar

1/2 dl (1/5 c) olive oil

Salt and pepper

1 bunch of green onions (ca. 6–8)

150 g (5 oz) celery

150 g (5 oz) ripe grape tomatoes

15 g (1/2 oz) dulse

Sautéed

30 g (1 oz) cashew nuts

150 g (5 oz) firm button mushrooms

15 g (1/2 oz) *nori* (laver)

200 g (7/8 c) bean purée

2 eggs

90 g (3/8 c) oatmeal

Fresh thyme leaves

Salt and pepper

A little butter and olive oil

Puréed

200 g (7/8 c) bean purée

Salt and cracked black pepper

1. Soak the beans for 8–10 hours.
2. Cook the soaked beans with salt, and possibly a little *konbu*, until they are done. Drain, reserving the cooking water.
3. Put 200 g (7/8 c) of the cooked beans aside in a bowl.
4. Purée 400 g (1 3/4 c) of the cooked beans with a little of the cooking water to achieve a firm consistency.
5. Set half of the purée aside in a small saucepan and cover it tightly.

Salad

1. Season the whole beans with the vinegar, oil, salt, and pepper. Slice the white portion of the green onions into thin rings. Cut the celery and tomatoes up into small cubes. Moisten the dulse and cut it into thin strips. Toss all of these with the beans. Check the seasonings and adjust to taste.

Sautéed

1. Toast the cashew nuts on a dry skillet and then chop them finely. Rinse the button mushrooms and chop them into small pieces. Cut the *nori* into small pieces.
2. Thoroughly mix the eggs, oatmeal, thyme leaves, and *nori* pieces into the bean purée. Stir in the chopped cashews and mushrooms. Season the mixture with salt and pepper and let it stand for about 1 hour.
3. Shape the mixture into small patties and fry them in a skillet in butter or oil until they are golden.

Puréed

Warm the bean purée that was set aside in the saucepan. Adjust the consistency with a little cooking water. Season with salt and pepper and sprinkle a little cracked black pepper on top.

Serving suggestions

Each of the preparations can be served on their own, as a side dish, or together to make a main meal.

> *Konbu, nori,* and dulse all contribute umami to these three ways of preparing the white beans.

Tomatoes in tomatoes, with mouthfeel

Serves 4

800 g (1 ¾ lb) tomatoes, all sizes, types, and degrees of ripeness

3 tbsp apple cider vinegar

Salt and pepper

200 g (7 oz) white onions

1 bunch green onions (ca. 6–8)

2–3 types of herbs, for example, basil, parsley, chives, tarragon, or cilantro

2–3 tbsp olive oil

50 g (ca. ¼ c) pumpkin seeds with soy sauce (see Flavour Accelerators)

Green peppercorns

Marinade

200 g (7 oz) overripe tomatoes

Salt and pepper

2 dl (⅘ c) rice vinegar

½ dl (⅕ c) fish sauce

½ dl (⅕ c) sugar

Juice from 1 organic lemon

Ground sea salt

Croutons

200 g (7 oz) stale bread

2 tbsp olive oil

1 clove of garlic

Salt and pepper

Crisp garlic

Many cloves of garlic

Olive oil

Tomatoes

1. Remove the stems and make a small cut in the top of the tomatoes. Blanch them for 15–20 seconds in boiling water. Then plunge them into ice cold water, peel them, and put them in a bowl with the apple cider vinegar, salt, and pepper.
2. Peel and trim the onions. Cut the white onions in half and make thin slices. Cut the white parts of the green onions into thin rings.
3. Chop the herbs coarsely.

Marinade

1. Heat a dry skillet and roast the unpeeled overripe tomatoes until they begin to change colour and are a bit charred. Season lightly with salt and pepper and place them in a blender. Add rice vinegar, fish sauce, and sugar and blend until the tomatoes are completely puréed. Fine-tune the taste with lemon juice and sea salt.
2. Let the marinade stand at room temperature.

Croutons

1. Rip the bread up into small, not too regular pieces. Crush the garlic with the back of a knife.
2. Heat up a skillet with a little olive oil and the garlic and toast the bread pieces in it until they change colour and are a little burnt at the edges.
3. Remove them from the skillet and place them on absorbent paper towels.
4. Season with salt and pepper.

Crisp garlic

1. Using a mini mandoline or a very sharp knife cut up the garlic cloves into very thin even slices.
2. Pour olive oil into a cold skillet and add the garlic slices. Heat them gently until they change colour and turn light brown. Pay attention because the difference between done and charred is a matter of seconds. Remove the garlic slices and lay them out on absorbent paper towels.

Serving suggestions

Distribute the tomatoes unevenly on an attractive platter. Sprinkle the onions and some green peppercorns on top and pour over the marinade. Top with the crunchy ingredients—crisp garlic, roasted pumpkin seeds, croutons—together with the herbs. Finally drizzle with a little olive oil. The idea is to present the tomatoes to resemble a 'carvery' by formally slicing the largest tomatoes with a carving set at the table, like a roast. The dish can be served as an entrée or a side dish.

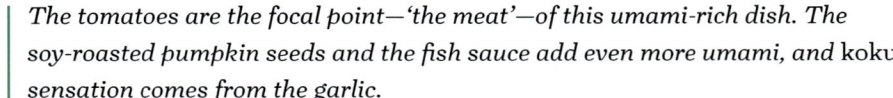

The tomatoes are the focal point—'the meat'—of this umami-rich dish. The soy-roasted pumpkin seeds and the fish sauce add even more umami, and koku *sensation comes from the garlic.*

Chilli-watermelon

Serves 4

1 ripe watermelon, ca. 3 kg (7-9 lb)
1 chilli pepper
2–3 tbsp rice vinegar
500 g (1 lb 1 ½ oz) skinless salmon fillet
1 tbsp sea salt
1 tbsp olive oil
4 shallots
2 tbsp neutral oil
Fresh cilantro

1. Cut the melon in half and then into 3 cm (2 ¼ in) thick slices. Cut off the peel.
2. Cut the melon slices to form cubes 3 cm × 3 cm (2 ¼ × 2 ¼ in). Set aside the pieces that were trimmed away.
3. Place the cubes in a container.
4. Cut the chilli pepper open, remove the seeds and membranes and cut it up into smaller pieces.
5. Blend the leftover melon with the chilli pieces and season with the vinegar to make a marinade.
6. Pour the marinade over the melon cubes and let them stand in a cool place for at least a couple of hours.
7. Lay the salmon fillet on a cutting board and score it deeply.
8. Rub sea salt thoroughly into the slits and bake it in an oven at 165°C (330°F) for 25 minutes.
9. Remove the salmon from the oven and pulse it in a blender.
10. Pour the olive oil into a skillet, turn the heat up to medium, and add the salmon. Toast the salmon by scraping it continuously from the bottom of the skillet with a spatula for about 25 minutes. When it is done it should resemble coarse breadcrumbs.
11. Transfer the dried salmon to paper towels to absorb any remaining fat.
12. Peel, trim, and slice the shallots. Place them in a skillet with cold neutral oil.
13. Heat them over medium heat until they turn light brown. Remove them and place on absorbent paper towels.

Serving suggestions

With a slotted spoon take the melon cubes out of the marinade and place them on a platter or individual plates. Sprinkle some of the dried salmon and sautéed onions on top. Garnish with fresh cilantro leaves.

The melon cubes will stay fresh and tasty for 4–5 days in the refrigerator. Only a little of the dried salmon is used in this recipe, but it is difficult to make a smaller portion of it. It can, however, be kept in the refrigerator or freezer for later use, for example, as a topping in salads or on vegetable dishes.

This very dry salmon is so full of umami and has such an intense taste that a teaspoonful is enough to make the watermelon taste much fuller. This dish, where the watermelon acts both as the vegetable and the central feature, is typical of Asian cuisine, although duck or pork fats would probably be used instead of olive oil. The chilli provides both heat and spice that work with the crisp, solid melon cubes and the intense umami in the dry salmon to create a filling dish that even a meat lover would find satisfying.

Bibliography

Allen, K. & J.W. Bennett. Tour of truffles: aromas, aphrodisiacs, adaptogens, and more. *Mycobiology* **49**, 201–212, 2021.

Andoh, E. *Kansha: Celebrating Japan's Vegan and Vegetarian Traditions*. Ten Speed Press, Berkeley, 2010.

Blumenthal, H., P. Barbot, N. Matsushisa & K. Mikuni. *Dashi and Umami: The Heart of Japanese Cuisine*. Cross Media Ltd., London, 2009.

Bozan, I. Pangenome analyses reveal impact of transposable elements and ploidy on the evolution of potato species. *Proc. Natl. Acad. Sci. USA* **120**:e2211117120, 2023.

Brunk, L. & P. Møller. Do children prefer colored plates? *Food. Qual. Pref.* **73**, 65–74, 2019.

Buckley, R. *Plants Taste Better*. Quarto Publ. Group plc, London, 2018.

Cardwell, G., J.F. Bornman, A.P. James & L.J. Black. A review of mushrooms as a potential source of dietary vitamin D. *Nutrients* **10**:1498, 2018.

Cheung, P.C.K. Mini-review on edible mushrooms as source of dietary fiber: preparation and health benefits. *Food Sci. Human Wellness* **2**, 162–166, 2013.

Chey, W.D., J. Kurlander & S. Eswaran. Irritable bowel syndrome: a clinical review. *JAMA* **313**, 949–958, 2015.

Cornish, M.L., A.T. Critchley & O.G. Mouritsen. A role for dietary macroalgae in the amelioration of certain risk factors associated with cardiovascular disease. *Phycologia* **54**, 649–666, 2015.

Cornish, M.L., A.T. Critchley & O.G. Mouritsen. Consumption of seaweeds and the human brain. *J. Appl. Phycol.* **29**, 2377–2398, 2017.

Cornish, M.L., O.G. Mouritsen & A.T. Critchley. A mini-review on the microbial continuum: consideration of a link between judicious consumption of a varied diet of macroalgae and human health and nutrition. *J. Oceanol. Limnol.* **37**, 790–805, 2019

Crawford, M.A. & D. Marsh. *The Shrinking Brain and the Global Mental Health Crises*. Authoritize, London, 2023.

Cwiertka, K. Popularizing a military diet in wartime and postwar Japan. *Asian Anthropology* **1**, 1–30, 2002.

Daverkosen, S., S. Ejlersen & O.G. Mouritsen. Progression towards an 80:20 (plant-based:animal-based) energy balance via specially designed Meal Kits. *Int. J. Food Design* **7**, 143–157, 2022.

Day, L. Proteins from land plants – potential resources for human nutrition and food security. *Trends Food Sci. Technol.* **32**, 25–42, 2013.

DeCosta, P., P. Møller, M.B. Frøst & A. Olsen. Changing children's eating behaviour – a review of experimental research. *Appetite* **113**, 327–357, 2017.

Dhingra, D., M. Michael, H. Rajput & R.T. Patil. Dietary fibre in foods: a review. *J. Food Sci. Technol.* **49**, 255–266, 2012.

Dunn, R. & M. Sanchez. *Delicious: The Evolution of Flavor and How It Made Us Human*. Princeton University Press, Princeton and Oxford, 2021.

Ergönül, P.G., I. Akata, F. Kalyoncu & B. Ergönül. Fatty acid compositions of six wild edible mushroom species. *Sci. World J.* **2013**:163964, 2013.

Esbaugh, W. H. History and exploitation of a serendipitous new crop discovery. In J. Janick & J. E. Simon (eds.) *New Crops*, pp. 132–139. John Wiley & Sons, New York, 1993.

Fabbri, A.D.T. & G.A. Crosby. A review on the impact of preparation and cooking on the nutritional quality of vegetables and legumes. *Int. J. Gastronom. Food Sci.* **3**, 2–11, 2016.

FAO, IFAD, UNICEF, WFP & WHO. *The State of Food Security and Nutrition in the World 2023: Urbanization, agrifood systems transformation and healthy diets across the rural-urban continuum.* Rome, FAO, 2023.

Fox, J. *On Vegetables*. Phaidon, New York, 2017.

Fuji, M. *The Enlightened Kitchen: Fresh Vegetable Dishes from the Temples of Japan*. Kodansha Intnl., Tokyo, 2005.

Griffith, C. & V. Valsamis. *The Vegetable*. Smith Street Books, Melbourne, 2017.

Grossman, A. Nutrient acquisition: the generation of bioactive vitamin B_{12} by microalgae. *Curr. Biol.* **26**, R319–R337, 2016.

Hachisu, N.S. *Preserving the Japanese Way: Traditions of Salting, Fermenting, and Pickling for the

Modern Kitchen. Andrews McMeel Publishing, LLC, Kansas City, 2015.

Hara, L. *The Japanese Larder.* Jaqui Small, London, 2018.

Hausner, H., A. Olsen & P. Møller. Mere exposure and flavour-flavour learning increase 2–3 year-old children's acceptance of a novel vegetable. *Appetite* **58**, 1152–1159, 2012.

Hawk, A. The great disease enemy, kak'ke (beriberi) and the imperial Japanese army. *Military Medicine* **171**, 333–339, 2006.

Hehemann, J.H., G. Correc, T. Barbeyron, W. Helbert, M. Czjzek & G. Michel. Transfer of carbohydrate-active enzymes from marine bacteria to Japanese gut microbiota. *Nature* **464**, 908–912, 2010.

Holdt, S.L. & S. Kraan. Bioactive compounds in seaweed: functional food applications and legislation. *J. Appl. Phycol.* **23**, 543–597, 2011.

Humphries, C. *Eat More Vegetables*. Penguin-Random House, London, 2016.

Japanese Culinary Academy. *Flavor and Seasonings: Dashi, Umami, and Fermented Foods*. Shuari Initiative Ltd., Tokyo, 2016.

Jønsson, S.R., S. Angka, K. Olsen, A. Tolver & A. Olsen. Repeated exposure to vegetable-enriched snack bars may increase children's liking for the bars – but not for the vegetables. *Appetite* **140**, 1–9, 2019.

Kalaras, M.D., J.P. Richie, A. Calcagnotto & R.B. Beelman. Mushrooms: A rich source of the antioxidants ergothioneine and glutathione. *Food Chem.* **233**, 429–433, 2017.

Kaoutari, A.E, F. Armougom, J.I. Gordon, D. Raoult & B. Henrissat. The abundance and variety of carbohydrate-active enzymes in the human gut microbiota. *Nat. Rev. Microbiol.* **11**, 497–504, 2013.

Kim, M.J, H.J. Son, Y. Kim, T. Misaka & M.R. Rhyu. Umami-bitter interactions: the suppression of bitterness by umami peptides via human bitter taste receptor. *Biochem. Biophys. Res. Commun.* **456**, 586–590, 2015.

Koeppel, D. *Banana: The Fate of the Fruit That Changed the World*. Hudson Street Press, London, 2008.

Kumudha, A. & R. Sarada. Effect of different extraction methods on vitamin B_{12} from blue green algae, *Spirulina platensis. Pharm. Anal. Acta* **6**:2, 2015.

Kushner, B. *Slurp! A Social and Culinary History of Ramen – Japan's Favorite Noodle Soup*. Global Oriental, Leiden, 2012.

Lane, M.L., E. Gamage, S.N. Ashtree, A.J. McGuinness, S. Gauci, P. Baker, M. Lawrence, C.M. Rebholz, B. Srour, M. Touvier, F.N. Jacka, A. O'Neil, T. Segasby & W. Marx. Ultra-processed food exposure and adverse health outcomes: umbrella review of epidemiological meta-analyses. *BMJ* **384**:e077310, 2024.

Makki, K., E.C. Deehan, J. Walter & F. Bäckhed. The impact of dietary fiber on gut microbiota in host health and disease. *Cell Host & Microbiota* **23**, 705–715, 2018.

Mann, C.C. *1493: Uncovering the New World Columbus Created*. Vintage Books, New York, 2012.

Marles, R.J. Mineral nutrient composition of vegetables, fruits and grains: the context of reports of apparent historical declines. *J. Food Comp. Anal.* **56**, 93–103, 2017.

Martin, L.E., K.E. Kay & A.-M. Torregrossa. Bitter-induced salivary proteins increase detection threshold of quinine, but not sucrose. *Chem. Sens.* **44**, 379–388, 2019.

McDougall, J. Plant foods have a complete amino acid composition. *Circulation* **105**:e197, 2002.

McGee, H. *On Food and Cooking: The Science and Lore of the Kitchen*. Scribner, New York, 2004.

McGee, H. *Nose Dive: A Field Guide to the World's Smells*. Peinguin Press, London, 2020.

Minich, D.M. & B.I. Brown. A review of dietary (phyto)nutrients for glutathione support. *Nutrients* **11**:2073, 2019.

Morton, O. *Eating the Sun*. Fourth Estate, London, 2007.

Mouritsen, O.G. *Seaweeds: Edible, Available & Sustainable*. Chicago University Press, Chicago, 2013.

Mouritsen, O.G. Gastrophysics of the oral cavity. *Curr. Pharm. Des.* **22**, 2195–2203, 2016.

Mouritsen, O.G. Deliciousness of food and a proper balance in fatty acid composition as means to improve human health and regulate food intake. *Flavour* **5**:1, 2016.

Mouritsen, O.G. The quest for umami. In T. Nishimura & M. Kuroda (eds.) *Koku in Food Science and Physiology*, pp. 33–46. Springer, Singapore, 2019.

Mouritsen, O.G. Roe gastronomy. *Int. J. Gastron. Food Sci.* **32**:100712, 2023.

Mouritsen, O.G. Time again for *meligarum*. *Nature Food* **4**, 530, 2023.

Mouritsen, O.G. When blue is green: seafoods for umamification of a sustainable plant-forward diet. *Int. J. Gastron. Food Sci.* **35**:100902, 2024.

Mouritsen, O.G. & C.V. Schmidt. A role for macroalgae and cephalopods in sustainable eating. *Front. Psychol*. **11**:1402, 2020.

Mouritsen, O.G. & K. Styrbæk. *Umami: Unlocking the Secrets of the Fifth Taste.* Columbia University Press, New York, 2014.

Mouritsen, O.G. & K. Styrbæk. *Mouthfeel: How Texture Makes Taste.* Columbia University Press, New York. 2017.

Mouritsen, O.G. & and K. Styrbæk. Design and 'umamification' of vegetables for sustainable eating. *Int. J. Food Design* **5**, 9–42, 2020.

Mouritsen, O.G. & K. Styrbæk. *Tsukemono: Decoding the Art and Science of Japanese Pickling.* Springer Nature Switzerland AG, 2021.

Mouritsen, O.G. & K. Styrbæk. *Octopuses, Squid & Cuttlefish: Seafood for Today and for the Future.* Springer Nature, Switzerland AG, 2021.

Mouritsen, O.G., L. Duelund, G. Calleja & M.B. Frøst. Flavour of fermented fish, insect, game, and pea sauces: garum revisited. *Int. J. Gastronomy Food Sci.* **9**, 16–28, 2017.

Mouritsen, O.G., L. Duelund, M.A. Petersen, A.L. Hartmann & M.B. Frøst. Umami taste, free amino acid composition, and volatile compounds of brown seaweeds. *J. Appl. Phycol.* **31**, 1213–1232, 2019.

Mouritsen, O.G., J.L. Pérez-Lloréns & P. Rhatigan. The rise of seaweed gastronomy: phycogastronomy. *Bot. Mar.* **62**, 195–209, 2019.

Mouritsen, O.G., P. Rhatigan, M.L. Cornish, A.T. Critchley & J.L. Pérez-Lloréns. Saved by seaweeds: phyconomic contributions in times of crisis. *J. Appl. Phycol.* **33**, 443-458, 2020.

Nishimura, T. & A. Egusa. "Koku" involved in food palatability: an overview of pioneering work and outstanding questions. *Kagaku to Seibutsu* **54**, 102–108, 2016.

Nowak, Z. *Truffle: A Global History.* Reaktion Books Ltd., London, 2015.

Olsen, A., C. Ritz, L. Kramer & P. Møller. Serving styles of raw snack vegetables. What do children want? *Appetite* **59**, 556–562, 2012.

Olsen, A., J.C. Sick, P. Møller & H. Hausner. No choice vs free choice: how serving situations influence pre-school children's vegetable intake. *Food. Qual. Pref.* **72**, 172–176, 2019.

Pérez-Lloréns, J.L. Microalgae: From staple foodstuff to avant-garde cuisine. *Int. J. Gastron. Food Sci.* **21**:100221, 2020.

Petroski, W. & D.M. Minich. Is there such a thing as "anti-nutrients"? A narrative review of perceived problematic plant compounds. *Nutrients* **12**:2929, 2020.

Petti, A., B. Palmieri, M. Vadalà & C. Laurino. Vegetarianism and veganism: not only benefits but also gaps. A review. *Prog. Nutr.* **19**, 229–242, 2017.

Poojary, M.M., V. Orlien, P. Passamonti & K. Olsen. Improved extraction methods for simultaneous recovery of umami compounds from six different mushrooms. *J. Food Comp. Anal.* **63**, 171–183, 2017.

Poore, J. & T. Nemecek. Reducing food's environmental impacts through producers and consumers. *Science* **360**, 987–922, 2018.

Provost, J., K.L. Colabroy, B.S. Kelly & M.A. Wallert. *The Science of Cooking: Understanding the Biology and Chemistry Behind Food and Cooking.* John Wiley & Sons, Hoboken, New Jersey, 2016.

Rico-Campà, A., M.A. Martínez-González, I. Alvarez-Alvarez, R. de Deus Mendonça, C. de la Fuente-Arrillaga, C. Gómez-Donoso & M. Bes-Rastrollo. Association between consumption of ultra-processed foods and all-cause mortality: SUN prospective cohort study. *BMJ* **365**:1949, 2019.

Saladino, D. *Eating to Extinction: The World's Rarest Foods and Why We Need to Save Them.* Farrar, Straus and Giroux, New York, 2022.

San Gabriel, A. & H. Uneyama. Amino acid sensing in the gastrointestinal tract. *Amino Acids* **45**, 451–461, 2013.

Sasano, T., S. Satoh-Kuriwada, N. Shoji, M. Iikubo,

M. Kawai, H. Uneyama & M. Sakamoto. Important role of umami taste sensitivity in oral and overall health. *Curr. Pharm. Des.* **20**, 2750–2754, 2014.

Schatzker, M. *The Dorito Effect: The Surprising New Truth About Food and Flavor.* Simon & Schuster, New York, 2015.

Schmidt, C.V. & O.G. Mouritsen. The solution to sustainable eating is not a one-way street. *Front. Psychol.* **11**:531, 2020.

Schmidt, C.V. & O.G. Mouritsen. Umami taste as a driver for sustainable eating. *Int. J. Food Design* **7**, 187-203, 2022.

Searchinger, T. *World Resources Report: Creating a Sustainable Food Future. A Menu of Solutions to Feed Nearly 10 Billion People by 2050.* World Resources Institute, Washington, 2019.

Simopoulos, A.P. The importance of the ratio of omega-6/omega-3 essential fatty acids. *Biomed. Pharmacotherapy* **56**, 365–379, 2002.

Smith, A.S. *Potato: A Global History.* Reaktion Books, London, 2011.

Solt, G. *The Untold Story of Ramen: How Political Crisis in Japan Spawned a Global Food Craze.* University of California Press, Berkeley and Los Angeles, California, 2014.

Srour, B., L.K. Fezeu, E. Kesse-Guyot, B. Allès, C. Méjean, R.M. Andrianasolo, E. Chazelas, M. Deschasaux, S. Hercberg, P. Galan, C.A. Monteiro, C. Julia & M. Touvier. Ultra-processed food intake and risk of cardiovascular disease: prospective cohort study. *BMJ* **365**:1451, 2019.

Tieman, D. *et al.* A chemical genetic roadmap to improved tomato flavor. *Science* **355**, 391-394, 2017.

Tree, I. *Wilding.* Picador, Pan Macmillan, London, 2018.

Vermeulen, S.J., B.M. Campbell & J.S.I. Ingram. Climate change and food systems. *Ann. Rev. Environ. Res.* **37**, 195-222, 2012.

Walton, S. *The Devil's Dinner: A Gastronomic and Cultural History of Chili Peppers.* St. Martin's Press, New York, 2018.

Willett, W. *et al.* Food in the Anthropocene: the EAT-Lancet Commission on healthy diets from sustainable food systems. *Lancet* **393**, 447–492, 2019.

Wobber, V., B. Hare & R. Wrangham. Great apes prefer cooked food. *J. Human Evol.* **55**, 340–348, 2008.

Woldegiorgis, A.Z., D. Abate, G.D. Haki, G.R. Ziegler & K.J. Harvatine. Fatty acid profile of wild and cultivated edible mushrooms collected from Ethiopia. *J. Nutr. Food Sci.* **5**:3, 2015.

Wrangham, R. *Catching Fire.* Basic Books, New York, 2009.

Yamaguchi, E. *The Well-Flavored Vegetable.* Kodansha International, Tokyo, 1988.

Zuckermann, L. *The Potato: How the Humble Spud Rescued the Western World.* North Point Press, New York, 1998.

Illustration credits

Unless specifically mentioned in the following all illustrations and photos are works of Jonas Drotner Mouritsen.

Ole G. Mouritsen, pp. xii, 25, 26, 27, 33, 40, 46, 63, 64, 78, 93, 111, 168; Klavs Styrbæk, pp. 21, 22, 23; Yann Fontana, p. 31; Kanga-an, p. 35; Shutterstock.com, p. 39 (mgstudyo), p. 42 (Splinex), p. 76 (Andris Tkacenko), p. 118 (Picture Partners), p. 119 (Luis Echeverri Urrea); Ajinomoto Co, Inc., p. 49; Kristoff Styrbæk, pp. 53, 84, 109, 193, 195, 203, 209, 269, 271; Dansk Tang, p. 67; Mathias Porsmose Clausen, p 87.

Index

A

acetic acid 121, 168
acidity 42, 60, 71–72, 76, 121
~ colour 72
acids 44, 50, 58, 72, 86, 107, 121–122. See also amino acids, fatty acids, nucleic acids
~ texture 84, 86, 88
acorns 28, 62, 107, 155
acorn squash 143
acrylamides 77
additives 52, 128–129
adenosine triphosphate 6
adenylate 45
Aframomum melegueta 160
aftertaste 83
aeging 66, 68, 77, 92, 94, 120, 148, 166, 168
agar 84, 106, 112, 114
Agaricus
~ *bisporus* 140
~ *campestris* 140
agriculture 11–13, 52, 99, 109, 114, 119, 125–127. See also farming
Agrocybe aegerita 60
aïoli 147, 158
air drying. See drying
Ajinomoto 48–49
aka-miso 158
aka-tosaka 78
Alaria esculenta 68, 78, 143
alcohol 50–51, 167
~ extraction 56
~ fermentation 29, 92, 156–158, 164, 168
~ non-alcoholic 7
aldehydes 50–51, 134, 137, 141
algae 17, 29. See also macroalgae, microalgae, seaweeds
alginate 8, 69, 84, 106, 112
alginic acid 112
alkaloids 110, 119, 141
allemande 59
allergies 113
allicin 153
alliin 153
alliinase 153
Allium
~ *ampeloprasum* 139
~ *cepa* spp. 140
~ *cepa* var. *aggregatum* 140
~ *fistulosum* 140
~ *sativum* 138
~ *schoenoprasum* 136
~ spp. 140, 151
all spice 160

almond milk 232
almonds 20, 28, 104, 116, 118–119, 226
alpha-linolenic acid 99, 104
Amaranthaceae 134, 143–144
amaranth family. See Amaranthaceae
Amaryllidaceae 136, 138–140
amaryllis family. See Amaryllidaceae
amazu-zuke 94
amino acids 45, 95, 99, 115. See also proteins
~ essential 98–99, 100
~ Maillard reactions 77
~ non-essential 99
AMP. See adenylate
amylase 106
amylopectin 106
amylose 106
Anaheim chilli 151
anandamide 65
anchovies 45, 83, 148, 152, 156, 164, 168, 238
anethole 137
animal welfare 15
anthocyanins 43, 58, 72–73, 110, 125, 132
Anthropocene 11–12
antibiotics 113
antinutrients 120
antioxidants 24, 28, 30, 71–73, 99, 110, 115, 121, 123, 125
ao-daimaru 133
ao-nori 34, 83, 166
ao-tosaka 78
Apiaceae 135–137, 140
Apicius 62
Apium
~ *graveolens* var. *dulce* 136
~ *graveolens* var. *rapaceum* 136
appetite 92, 122–123
apples 28, 50–51, 76, 112, 132, 182, 242
~ cider 168
~ ripeness 58–59
~ structure 44, 59, 88
apricots 119
aquafaba 136, 158
arachidonic acid 30, 104
archaea 112
Armoracia rusticana 138
aroma 1–2, 50–51, 91, 99. See also aromatics
~ browning 76–77
~ evolution 50
~ families 50

~ fermentation 95, 122
~ genes 52–53, 119
~ mushrooms 60
~ plants 44, 50–51, 131
~ profile 131
~ seaweeds 66
~ tomatoes 52–53
~ truffles 62, 64–65
~ vegetables 24, 47, 50–51, 92, 94
aromatics 7, 50–52
Arthrospira
~ *maxima* 30
~ *platensis* 30
artichokes 19, 118, 132
arugula 132
asa-zuke 92
ascorbic acid 76, 115
Asparagaceae 132
asparagus 18–19, 72, 87, 90–91, 132, 220, 224, 226
~ family. See Asparagaceae
~ nutrients 115
~ taste and aroma 44, 132
asparagusic acid 132
Asparagus officinalis 132
Aspergillus oryzae 95, 156
Asteraceae 132, 134, 138, 142
aster family. See Asteraceae
astringency 42–44, 58, 76, 119
atherosclerosis 105
ATP 6
aubergines 24, 77, 133, 206, 210, 212, 214, 216, 218
~ taste and aroma 133
Australopithecines 5
autoimmune diseases 113
avocados 28, 100, 104, 115–116, 118, 133
Aztecs 30
azuki beans 133
AA. See arachidonic acid

B

bacon 58, 68, 148
bacteria 30, 44, 113. See also cyanobacteria, lactic acid bacteria, microorganisms
~ intestinal 112–114
~ pathogenic 121
bagna càuda 148
baker's yeast 159
balsamic vinegar 77, 148, 167–168, 184
banana peppers 57
bananas 38–39, 42, 58, 107,

112, 118
barbeque sauce 59
barley 95, 102, 156, 158, 166
basic tastes 44
basil 18, 161, 230
beans 72, 107, 133, 176, 248
~ cooking 71, 88, 133
~ fermented 45, 95
~ nutrients 112, 118
~ proteins 100
~ soaking 133
~ taste and aroma 133
béarnaise sauce 158
béchamel 59
beef 47, 102, 104, 148–149
beefsteak tomatoes 144
beer 7, 47, 76–77, 157, 164
beets 19, 73–74, 88, 134. See also red beets
~ nutrients 58, 112, 115, 118
~ taste and aroma 51
Belgian endives 134
bell peppers 19, 51, 57, 83, 116, 134, 240
beni-shoga 94
bergamot 151
berries 6, 58, 115, 121
beta-carotene 73, 115
beta-damascenone 144
beta-ionone 144
betalains 71, 73, 134
Beta vulgaris 144
~ convar. *cicla* 144
~ convar. *crassa* var. *conditiva* 134
betaxanthins 73, 134
beverages 7
bile salts 105
bioavailability 121–123
biodiversity 12–13, 126
bitter 36, 43, 50, 66, 119
bitterness 90, 132, 134–135, 141
~ increase 125
~ personal perception 7
~ reduction 45, 47, 83, 95, 180
black garlic 46, 77, 149
black pepper. See pepper
black salsify 19, 134, 226
bladder 68
bladderwrack 19, 66, 69, 78, 143
blanching 44, 71–72, 76, 78
Bleu de Bresse 149
bloating 114
blood sugar 106–107, 112
blueberries 72, 115
blue-green microalgae 30.

276 Index

See also cyanobacteria
blue mussels 69, 165
blue-veined cheeses *149*, 256
boiling 41, 72, 88, 91, 110, 162
bok choy *134*
Boletus edulis 45, 60, 115, 140
Bolognese sauce 59, 166
bonito 32, 152, 156
borage 28
borlotti beans 133
botargo 149
bottarga 149
bouillon *149–150*. *See also* soups, stocks
bouquet garni 136
brain 52, 122
brain, human 5–6, 30, 41, 104–105, 113, 116
brain-stomach axis 122
Brassica
~*juncea* 159
~*napus* ssp. *rapifera* 139, 142
~*napus* var. *pabularia* 142
~*nigra* 159
~*oleracea* 142
~*oleracea* convar. *acephala* 139
~*oleracea* convar. *capitata* var. *alba* 135
~*oleracea* convar. *capitata* var. *conica)* 135
~*oleracea* convar. *capitata* var. *rubra* 135
~*oleracea* convar. *capitata* var. *sabauda* 135
~*oleracea* convar. *fruticosa* var. *gemmifera* 135
~*oleraceae* convar. *botrytis* var. *botrytis* 136
~*oleraceae* convar. *botrytis* var. *italica* 135
~*oleracea* var. *alboglabra* 135, 138
~*oleracea* var. *gongylodes* 139
~*oleracea* var. *palmifolia* 139
~*oleracea* var. *viridis* 137
~*oleracea* var. *viridis* × *gemmifera* 139
~*rapa* ssp. *pekinensis* 136
~*rapa* ssp. *rapa* 142, 144
~*rapa* var. *chinensis* 134
Brassicaceae 132, 134–136, 138–139, 142, 144
Brazil nuts 118
bread 150. *See also* croutons
~-crumbs 150, 160, 164
breast milk 6, 14
breeding. *See* plant breeding
Brillat-Savarin, Jean Anthelme 62
broccoli 1, 19, 28, *135*, 222. *See also* broccolini

~ browning 77
~ nutrients 116, 118
~ taste and aroma 7, 44, 54, 95, 135
~ texture 90
broccolini 28, 95, *135*, 178
bromine 51, 66
broth 149. *See also* bouillon, *dashi*, soups
brown beans 118
browning 76–77
Brussels sprouts 7, 54, 116, 118, *135*, 252
buckwheat 100
Buddhism 32, 34, 83
bulbs 18–20, 22, 46, 58
burning 42–43, 56, 151, 154, 159–160
burning/irritating taste 43. *See also* capsaicin, isothiocyanate, piperine
butter 58–59, 110
butternut squash 143
buttery 76
button mushrooms 19, 51, 140
butyric acid 160

C

cabbage family. *See* Brassicaceae
cabbages 19, 24, 51, 83, 90, *135*, 224, 226
~ browning 76–77
~ kimchi 179
~ nutrients 112, 115
~ taste and aroma 50, 54, 95
Caesar salad 83
caffeine 119
calcium 113, 118–119, 121
~ carbonate 119
~ channels 47
~ citrate 86
~ ions 86, 88
Calçotada Festival 20–23
calçots 20–23
calendula 28
calories 6, 12, 30, 103, 106, 119. *See also* energy
~ low 90, 95, 100, 112, 118
~ potatoes 108–110
Calvatia gigantea 60, 140
camphor 43, 137
cancer 12, 113–114, 119
candy cane beets 74
Cantharellus cibarius 140
caper berries *150*
capers *150*, 158, 164, 220, 240
Capparis spinosa 150
capsaicin 43, 54, 56–57
capsaicinoids 56, 151

capsicum 134
Capsicum annuum 134
Capsicum spp. 151
caramel 76
caramelisation 76–77, 149, 151, 154
caraway 58
carbohydrates 8, 30, 58, 72, 86, 106–107, 121. *See also* dietary fibre, polysaccharides, starch, sugar
~ breakdown 29, 95, 106
~ contents 30, 107, 123
~ diet 97, 99, 103, 107, 119
~ dietary fibres 84, 86, 112–114
~ energy 74, 106–107, 110
~ fermentation 29, 95
~ gelatinisation 6
~ insoluble 8
~ water binding 58, 86
carbon 12
~ dioxide 12, 72, 126
~ footprint 15
cardiovascular diseases 1, 30, 99, 104
Carême, Marie-Antoine 18, 59
carnivores 98, 120
carnosine 98
carob powder 84
Carolina Reaper 57
carotenoids 30, 53, 71, 73, 110, 115, 125
carrageen 79
carrageenan 8, 84, 106, 112
carrots 18–19, 44, 73–74, 83, 87, 88, *135*, 228
~ browning 76–77
~ nutrients 58, 112, 115–116, 118
~ taste and aroma 135
caryophyllene 133
cashew nuts 162–163
cassava 13, 119, 122
Catherine de Medici 62
cauliflowers 19, 28, 72, 77, *136*, 252, 255
~ nutrients 112, 116, 118
~ taste and aroma 54, 136
Cavendish, William 38
celeriac 18–19, 76, 118–119, *136*, 182, 200
celery 18–19, *136*, 202
~ seeds 182
cell membranes 50, 86–88, 105
cell network 87
cellulose 8, 58, 84, 86, 88, 90, 106, 112
cell walls 8, 30, 58–59, 84, 86–88, 90, 105–106

cephalopods 2, 99, *150*
cereals 34, 47, 100, 107, 112, 166. *See also* barley, grains, oat, wheat
chaconine 110
cha-kaiseki 36
chanterelles 140, 158
chavicine 161
cheese 149, 184, 190, 198, 200. *See also* blue-veined cheese, Parmesan cheese
~ fats 104
~ flavour 2, 83
~ fresh 246
~ goat 244, 250
~ *koku* 45, 47, 49
~ texture 84
~ umami 45, 47
chemesthesis 42–43, 151, 161
cherries 115, 119, 236
cherry peppers 57
cherry tomatoes 144
chestnut mushroom 60
chestnuts 28
chewing 6, 13, 122–123
chia seeds 100, 104
chicken 47, 91
chickpeas 28, 100, 118, *136*, 158
chicory 134
children 1–2, 13–15, 95, 116, 194
chilli peppers 19, 24, 34, 42–43, 54, 56–57, 83, *151*, 178, 186, 242, 250
Chinese broccoli 135, 138
Chinese cabbage 54, *136*
Chinese radish 94, 137. *See also* daikon
Chioggia beets 74
chitin 8, 60, 106, 112
chives *136*
Chlorella 30, 120
chlorophyll 18, 30, 71–72, 78, 90
chloroplasts 78, 86, 132
chocolaty 77
cholecalciferol 98
cholesterol 104–105, 110, 112
Chondrus crispus 78–79
choriceros 20, 163
chrome 118
Cicer arietinum 136
Cichorium intybus var. *foliosum* 134, 142
cinnamaldehyde 50
cinnamon 50, 119
citric acid 44, 121, 142, 151
Citrullus lanatus 140
citrus fruits *151*
Citrus junus 168
Citrus spp. 151

Index 277

climate change 1, 8, 11–12, 125
cloves 50
cobalamin 115–116
cod 255
coffee 77, 119
collagen 149
collard greens *137*
Colorado potato beetle 109
colorectal cancer 114
Columbus 56
common mushrooms 140
condiments 1, 147, 153, 158–159, 162–163, 167
cone cabbage 54, 135, 226
confit 151
conserving 92, 121, 153, 157. See also pickling, fermenting, preserving
constipation 114
convar. 132
conventional farming 2, 125–127
Convolvulaceae 144
cooking in evolution 6
copper 113, 118
corn 8, 13, 19, 45, 52, 76–77, 104, 110, *137*
~ oil 116
~ starch 160
cornichons 137, 162
coumarins 119
courgettes 145
couscous 252
crab shells 165
cranial nerves 43
Craterellus cornucopioides 140
craving 52, 57
creamy 7, 43, 76
creatine 98
cremini 51, 140
CRISPR 39
crispy 1, 7, 59, 84, 86–87, 92
croutons 58, 84, 150
crudités 135–136, 142, 147
crunchy 1, 7, 43, 58, 87, 92, 123
cucumbers 28, 50, 51, 84, 94, *137*, 162, 172, 186
Cucumis anguria 137
Cucumis melo ssp. 140
Cucumis sativus 137
Cucurbita
~ *maxima* Duchesne 143
~ *maxima* var. *kabocha* 143
~ *moschata* 143
~ *pepo* 145
~ *pepo* var. *turbinata* 143
Cucurbitaceae 137, 140, 143, 145
cucurbitacin 137
culinary skills 129
cultivar 132

Curcuma longa 138
curly parsley 140
cuttlefish 150
cyanide 119
cyanidin 90
cyanobacteria 30, 116
Cynara cardunculus var. *scolymus* 132
cysteine 77, 115

D

daikon 34, 87, 94, *137*, 173, 234
dairy products 2, 13, 15, 47, 58, 83–84, 98, 100, 102, 110, 116, 119–120, 171
Danablu 149
dandelion greens 18
dashi 32–34, 83, 103, 149, *152*, 156, 165
~ *katsuobushi* 156
~ mushrooms 60
~ *ponzu* and *sanbaizu* 162
~ principle 48
~ seaweeds 66
~ *shiitake* 34
~ *tsukemono* 167
~ vegan 32, 152
Daucus carota ssp. *sativus* 135
daughter sauces 59
dehydrating 60, 84, 143, 150, 158. See also drying
~ vegetables 92, 94, 121
deliciousness 5, 15, 29, 34, 41, 47, 50, 83, 122, 171
~ umami 34, 102
De re coquinaria 62
detoxification 119
DHA. See docosahexanoic acid
diabetes 1, 12
diet 11–12, 15, 73, 99, 125
~ complete 100
~ evolution 50
~ fatty acids 105
~ fibres 113–114
~ healthy. See healthy diet
~ Japanese 102–103, 118
~ meat 6
~ planetary 13
~ plant-rich 1–2, 7–8, 41, 127, 171
~ vegan and vegetarian 98, 116, 120
digestion 6, 13, 30, 73, 86, 95, 98, 106, 112, 114, 121–122. See also enzymes
digestive system 6, 8, 97, 100, 107, 113, 116, 119–120, 122–123
dimethyl sulfide 51, 65–66,

132, 137, 144
dinosaur kale 139
Dioscorea spp. 144–145
dipeptides 47
disaccharides 106
diseases 1, 12, 97, 128
DNA 115–116
docosahexaenoic acid 30, 98–99, 104
donko 60
Dorito effect, the 52
drying 41. See also dehydrating
~ mushrooms 60
~ seaweeds and algae 30, 66, 68
~ seeds 28
~ vegetables 87, 94, 167
dulse 19, 51, 78, 105, 120, 143, 254
Dunn, Rob 50

E

earthy 51, 60, 65, 77
EAT-Lancet Commission 12–13, 15, 116
E. coli 121
ecological 1, 99, 126, 128
eel 34
eggplants. See aubergines
eggs 2, 13, 59, 64, 83, 98, 103–104, 116, 120, 158, 171, 196, 200
eicosapentaenoic acid 30, 104
ekitai shio-koji 156
Elettaria cardamomum 138
emulsion 161, 168
energy 11, 15, 126
~ depots 8, 44, 74, 86
energy of food 6, 18, 30, 98–100, 104, 106–107, 112
Engraulis encrasicolus 148
environment 1, 11–13, 15, 125–126
enzymes 44, 48, 50, 72, 99, 104–105, 115, 120, 122–123
~ browning 60, 72, 74, 76
~ digestive 8, 86, 100, 106, 112, 119, 122–124, 152
~ firming 88, 90
~ garlic 46
~ *koji* 95, 156
~ microorganisms 2, 112–114, 166
~ predigestion 122
~ ripening 59, 87
~ taste and aroma release 29, 44, 47, 50, 54, 60, 106, 135, 141, 149, 153, 159, 164
~ tenderising 68, 84

EPA. See eicosapentaenoic acid
ergosterol 106
Eruca vesicaria ssp. *sativa* 132
Escoffier, August 59
espagnole 59
esters 50–51
estragol 137
etiolation 90
eugenol 50
evolution, human 2, 5–6, 11, 13, 41, 50, 98–99, 105, 120

F

Fabaceae 133, 136, 140, 141
famine 103, 108–109, 126
farming 11, 125–127. See also agriculture
fatigue 113
fats 8, 29, 30, 58, 60, 68, 104–106, 115. See also cholesterol, fatty acids
~ blood content 112
~ contents 47, 86, 98, 104, 110
~ diet 97–99, 102–103, 107, 114, 120–121
~ essential 104
~ low 68, 110, 118
~ solubility 73
fatty acids 50, 99, 104–106, 113. See also omega-3, omega-6
fava beans 116, 133
fenchone 137
fennel 19, 58, 107, *137*, 155
fermenting 2, 41, 45, 49, 92, 106, 116, 119, 121. See also enzymes, *koji*, microorganisms, yeast
~ alcohol 29
~ beverages 7, 164
~ garlic 46
~ legumes 45, 158
~ medium 94
~ nutritional value 122
~ sauces 81, 152–153, 166
~ taste and aroma 47, 83, 122
~ texture 84, 122
~ vegetables 32, 95, 116, 121–122, 167
~ vinegar 168
fertilisers 12–13, 52, 109, 126–127
feta 240
fibre 8, 24, 90
~ dietary 84, 86, 99, 106, 108, 110, 112–114, 123
~ digestion 113
~ insoluble 86, 112
~ microbiota 112–114
~ soluble 86, 112

278   Index

~ texture 84, 88, 106
Ficus carica 138
field mushrooms 140
figs *137*, 198
fire and cooking 6
fish 2, 5, 13, 34, 78, 102–103, 110, 116, 120, 171, 180, 255
~ dried 32, 60
~ fatty acids 104–105, 121
~ fermented 32, 95
~ flakes *152*
~ nutrients 99–100
~ populations 12
~ roe 2, 149, 260
~ sauce 45, 83, *152*
fisheries 12
flageolet beans 133
flavonoids 71–72, 125
flavour 50. *See also* aroma, taste
~ accelerators 82
flax seeds 104, 118, 155, 208
flexitarian 2, 83, 129
floral 51, 77
flowers 17–19, 28, 51, 135
fly agaric 120
Foeniculum vulgare var. *azoricum* 137
folate 143
folic acid 115
food culture 1, 5, 13, 24, 45, 56, 98, 102
food hygiene 121
food preferences 5–6, 102, 129
food production, global 126
food waste 12–13, 32, 71, 83, 129, 171
food web 31, 104–105
foraging 60, 62
Fragaria spp. 143
Francis I of France, King 62
Frederick the Great 108
French cuisine 18, 59
French fries 156, 159, 162
French Revolution 109
fricassee 204, 224
fructose 76–77, 106
fructus 17
fruiting bodies 7–8, 29, 60, 62, 65, 74
fruits 6–7, 13–14, 17–19, 28, 120
~ colours 72–73
~ nutrients 99–100, 104, 106–107, 112, 115, 118
~ poisons 119
~ taste and aroma 6–7, 42, 44–45, 50, 58–59, 83
~ texture 84, 86–88, 121
~ umami 45
~ unripe 7, 44

fruit sugar. *See* fructose
fruit vinegar marinade *152*
fu 34
fucha ryōri 34
fucosterol 106
fucoxanthin 78, 115
Fucus serratus 78
Fucus vesiculosus 78
~ 143
fulness 122
fungi 2, 17, 29, 44–45, 106, 112, 115, 119, 125
fungicides 38, 99, 110, 125–126
funori 78
furfurylthiol 155, 166
furikake 69, *153*
furu-zuke 92
Fusarium oxysporum 38
Fushimi chilli peppers 24

G
gai lan 135, *138*
galactan 106, 112
galactose 106
gamma-Glu-Cys-Gly 47
gamma-Glu-Val-Gly 47
garden peas 141
garden tomatoes 144
gari 138, 154
garlic 18, 36, *138*, 147, *153*, 268
~ black. *See* black garlic
~ *koku* and *kokumi* 45–47
~ nutrients 115–116
~ taste and aroma 153
~ wild 18
garum 152–153, 168
gastriques 154
gastrointestinal tract 112, 120
gelatine 58
gelatinisation 6, 8, 86, 90
gelation agents 8, 69, 112, 131
gene bank 108
genus 131
geosmin 51, 134
gherkins 137, 162
ginger 18–19, 34, 83, 94, *138*, 154, 166, 186, 212
~ family. *See* Zingiberaceae
~ pickled 154
~ sensory perception 54, 138
gingerol 138, 154
GLA. *See* linolenic acid
Gloiopeltis spp. 78
glory family. *See* Convolvulaceae
glucan 8, 106, 112
glucose 77, 86, 90, 106, 156, 158
~ glycaemic index 107
glucosinolates 135

glutamate 66, 115, 149, 157–160, 166
~ MSG 48
~ receptors 122
~ umami 45, 60, 66, 99
glutamic acid 45, 99. *See also* glutamate
glutathione 45–47, 60, 115
glycaemic index 28, 107, 110
Glycine max 133
glycogen 8
GMO 39
GMP. *See* guanylate
goma 116, 155
~ -shio 155
Gorgonzola 149
Gouda cheese 47
gourd family. *See* Cucurbitaceae
Gracilaria spp. 78
grains 13–15, 95, 100, 107, 118. *See also* barley, rice, wheat
~ sprouted 120, 200
grains of paradise 160
gram lentils 140
grapefruit 151
grapes 42, 58, 167–168
grape sugar. *See* glucose
grape tomatoes 144
grass family. *See* Poaceae
gratin 214
Great Food Transformation, The 12
green beans 28, 72, 115–116, 176, 204, 248
green cabbage 54, 135
green cuisine 1–2, 17, 83
greenhouse gases 12, 126
green onions 140
green peas 45
green peppercorns 160
Green Revolution, The 126
greens 18
green salad 143
greens to go 83–84
green string lettuce 143
gremolata 155
Gros Michel 38
guanylate 45, 60, 152, 158
guar gum 84
gulfweed 120
gum arabic 84
gyoza 103

H
habñeros 151
haem iron 98
halogenated compounds 51
ham 45, 83, *155*, 238
Hass variety 133
hazelnuts 20, 28, 162

head lettuce 142
healthy diet 1–2, 12, 15, 52, 97, 102–103, 105, 112, 116, 120, 125
heart disease 12
heavy metals 99
hedonic taste 42
Helianthus tuberosus 138
hemicellulose 8, 58, 84, 86, 88, 136
hemp seeds 34, 83, 100, 104, 155, 162, 166
hepatopancreas 152
herbicides 99
herbivores 5–6
herbs 1–2, 7, 18, 28, 34, 102, 104, 180
hexanal 144
hibernation 44
Higuchi, Masataka 24
Higuchi-Noun 24–27
hijiki 120
Hinduism 32
Hippocrates 97
histidine 99
Hizikia fusiformis 120
Hoisin sauce 155
Hokkaido squash 143
hollandaise sauce 59, 158
homeostatic mechanism 122
hominins 5–6
Homo erectus 5
Homo sapiens 5, 98, 120
hon-dashi 152
hormones 97, 99, 105
horn of plenty 140
horseradish 42–43, 50, 54, *138*, 202, 220
hoso-maki 69
hot 54, 56–57, 151, 160, 164
hummus 136, 141, 263
hunger 11–12, 109
hybrids 26, 38, 52, 57, 131
hydrazines 120
hydrogels 58
hygiene 121
hypertension 12

I
Iberian black foot pigs 155
iceberg lettuce 142
Ikeda, Kikunae 48
ikitai shio-koji 95
immune system 113, 122
IMP. *See* inosinate
Inca culture 108
industrialisation 11, 52, 64
industrialised foods 15, 52, 113, 128
Industrial Revolution, The 126

inflammatory conditions 104–105, 113–114
inosinate 45, 148–149, 152, 156, 165
insecticides 99
insects 2, 8, 95, 127, 153
insulin 106–107
International Potato Centre, The 108
intestines 6, 97, 106, 112–116, 123
inulin 8, 106, 112, 138
iodine 51, 66, 118, 120–121
Ipomoea batatas 144
Irish moss 78–79
Irish Potato Famine 109
iron 98, 108, 118, 121
irritating 42–43, 54, 151
isoleucine 99
isothiocyanates 43, 50–51, 135, 138, 142, 159

J

Jaffrey, Madhur 57
Jainism 32
jalapeños 57, 151
jamón ibérico 238
Japan 1, 24, 32, 48, 83, 92, 95, 102–103, 116, 118
Japanese cuisine 48, 78, 92, 152–153, 162, 164
jelly 86
Jerusalem artichokes *138*
juicy 58–59, 84, 87, 212

K

kabocha squash 143
kainic acid 120
kaiseki 36
kale 14, 69, 72, *139*, 196, 242
~ nutrients 116, 118
~ sprouts 139
~ taste and aroma 54, 139
kalettes *139*, 234
Kamo aubergines 24
Kampot pepper 161, 180, 204
karashi-zuke 94
kasu-zuke 94, 123
katsuobushi 32, 34, 60, 83, 150, 152, *156*, 162
kernels. See nuts, seeds
ketchup 156
kidney beans 28, 104, 118–119, 133, 180, 236
kimchi 116, 179
kiwi 112
knotweed family. See Polygonaceae
koe-chiap 156
kohlrabis 19, 73, 90, *139*, 142, 173
koji 95, *156*, 158, 164. See also shio-koji
~ -kin 95
~ soy sauce 166
~ vegetables 92, 95, 178, 180
~ -zuke 94
kokabu 26
koku 29, 45–49, 60, 65, 83, 149, 152, 164
~ enhancing salty and sweet 47
~ garlic 138, 147
~ reducing bitterness 47
~ umami 47
kokumi 45–46, 48–49
kombucha 7
konbu 19, 66, 83, 118, 120, 143, 165
~ *dashi* 152
~ glutamate 48
~ *tsukedani* 68–69, 165
~ umami synergy 32
Kuho spring onions 24
Kyoto vegetables 24–27, 133
kyō-yasai 24–27
kyuri asa-zuke 94

L

lacinato kale *139*
lactic acid 121
~ bacteria 92, 94, 116, 122, 162, 179
lactose 106
Lactuca
~ *sativa* convar. 142
~ *sativa* var. *capita* 142
~ *sativa* var. *capitata crispum* 142
~ *sativa* var. *crispa* 142
~ *sativa* var. *longifolia* 142
Lake Chad 30
Laminaria digitata 164
Laminariales 118, 120
lasagna, green 218
Lauraceae 133
laurel family. See Lauraceae
lavender 28
laver 66, 114, 143
leaf lettuce 142
leaves 7, 18–19, 44, 51, 58, 72, 90, 104, 121
lectin 119, 133, 140
leeks 18–20, 36, 58, 101, 116, *139*, 149, 240
lees 157. See also sake kazu
legumes 13, 19, 28, 34, 100, 118, 120. See also beans, lentils, peas, pulses
lemon 72, 76, 88, 151, 155, 174, 182

Lens
~ *culinaris* 140
~ *esculenta* 140
~ *esculenta puyensis* 140
Lent 18
lenthionine 51, 60
lentils 28, 100, 112, 116, 118, *140*, 264
Lentinus edodes 60, 140
lettuces 18–19, 58, 142
leucine 99
lifestyle diseases 52
lignin 8, 90, 106, 112
lima beans 118–119, 133
lime 151
lime-soy-fish sauce dressing *157*
limonen 50, 137, 151
linalool 138
linoleic acid 104, 137
linolenic acid 30, 99, 104, 137
lipoproteins 105
lipoxygenase 50
Loligo forbesii 150
longevity 118
lotus root 19, 34
Louis XVI 109
Lumpers 109
lupini beans 133
lutein 110
lysine 77, 99, 100

M

macroalgae 143. See also seaweeds
Macrocystis pyrifera 66
macronutrients 97, 99–100, 123
Madagascar peppercorns 160
magnesium 72, 118
~ ions 86, 88
Maillard
~ compounds 151, 166
~ reactions 46, 76–77, 149
maize. See corn
malic acid 44, 121, 142
malnutrition 12, 30, 128
maltose 77, 106, 144
malt sugar. See maltose
Malus domestica 132
Malus spp. 132
manganese 118
mango 58, 73
Manihot esculenta 119
mannitol 69
Marie-Antoinette 109
marinade. See marinating
marinara 59
marinating 41, 68, 92, 150, 156–157, 162, 168. See also pickling

~ *dashi* 152
~ fish sauce 152–154
~ mushrooms 157
~ sauces 158, 166
~ vegetables 58, 84, 87, 92–94, 152, 154, 167, 172–173, 178–180
Marmite *157*, 159
marrows 19, 28, 145
matsutake 62
Mauléon, Paul 62
mayonnaise 136, 147, *158*, 162, 200, 222
McGee, Harold 50
mealy 7–8, 44, 58–59, 87–88, 141
meat 2, 6–7, 13, 15, 18, 45, 47, 83, 95, 100, 104, 113–114, 116, 120, 149, 155, 171
~ eating 5, 7, 32, 34, 102–103
~ fake 52
~ vs plants 5
meaty 77
Meiji period 24, 102
meligarum 153–154
melons 50–51, 73, 94, *140*, 270
mental health 36, 97, 105–106, 113
menthol 43, 50
Meristotheca palulose 78
metabolism 97, 114
methionine 99–100
methylbutanal 144
methylisocyanate 150
methylthiopentanenitrile 135
microalgae 29–30, 104–105, 116, 120
microbiota 97–98, 112–114. See also microorganisms
micronutrients 97, 108, 118, 121
microorganisms. See bacteria, fungi, microalgae, yeast
milk 59, 104, 160
~ sugar. See lactose
mindfulness 36
minerals 72, 86, 97, 110, 112–113, 118, 121–123
mint 50, 153, 230, 240
mirin 34, 83, *158*
miso 34, 83, 92, 95, *158*, 214
~ *dengaku* 210
~ -mayonnaise 200, 222
~ soup 102, 158, 188
~ -zuke 94
mitochondria 115
mizuna 24
modoki ryōri 34
molluscs 2, 152
molybdenum 118
monk cuisines 32–35

monoculture 13, 38, 52, 109, 125–126
monosaccharides 106. *See also* fructose, galactose, glucose
Moraceae 138
mossy 51
mouthfeel 8, 42, 54, 58. *See also* texture
MSG 48, 160. *See also* glutamate
mucous membranes 42–43, 54, 58, 114
mugi-miso 158
multisensory experience 1, 7, 41, 66
mung beans 133, 140, 192
muscles 6, 98
mushrooms 1–2, 7–8, 17, 19, 29, 34, 125, *140*, 218, 238, 246, 258, 262
~ browning 76
~ colours 74
~ marinated 157
~ nutrients 98–100, 104–107, 112, 115–116, 118
~ poisons 119–120
~ powder 158
~ taste and aroma 34, 41, 45, 51, 60, 140
muskmelons 140
mussels 69, 165. *See also* scallops
mustard 42–43, 50, 54, 94, 159
~ family. *See* Brassicaceae
~ seeds 159, 182
mycelium 29, 62, 65
mycorrhiza 62, 64
myristicin 155
Mytilus edulis 165

N

napa cabbage 136
nasturtium 28
nasu dengaku 210
neophobia 13–14
Nereocystis luetkeana 66
nerve cells 8, 42–43, 116
neurodegenerative diseases 105
neurotoxins 119–120
nightshade family. *See* Solanaceae
nitrogen fixation 127
nondienal 137
nori 66, 69, 114, 143
nucleic acids 45, 77, 95, 122
nucleotides 6, 45, 60, 95, 122. *See also* adenylate, guanylate, inosinate

nuka-zuke 94, 123
nutrition 1–2, 11, 14, 97, 103, 128
~ organic 125, 127
nutritional value 6–7, 13, 30, 52, 71, 92, 105, 107, 110, 122
nutritional yeast 159, 208
nuts 13, 15, 17, 28, 58, 84, 120, *162–163*
~ contents 100, 104, 112, 115–116, 118
nutty 76

O

oarweed 164
oat 115–116
Obaku 34
obesity 1, 52, 113–114
octanol 60
octopuses 150
Octopus vulgaris 151
ogonori 78
oleic acid 28
omega-3 30, 98–99, 104, 121, 125, 133
omega-3/omega-6 balance 105
omega-6 30, 99, 104
omega-6/omega-3 balance 105
omnivores 5, 14, 98
onions 18–20, 24, 44, 58, 72, *140*, 190, 238
~ browning 76
~ charred 151
~ nutrients 115
~ taste and aroma 36, 50–51, 54
oranges 45, 151
oregano 18, 50, 226
organic farming 125–127
organoleptic test 56
osmotic pressure 87
oxalic acid 44, 119, 142–143
oxidation 76, 115. *See also* antioxidants
oyster mushrooms 29, 140
oyster sauce 160

P

pain 8, 43, 54, 161
Palmaria palmata 66, 68, 78, 105, 120, 143
Panama disease 38
panko 150, *160*
papaya 115
Paris Green 109
Parmentier, Antoine-Augustin 108
Parmesan cheese 45, 47, 83, *160*
Parmigiano-Reggiano 160
parsley 116, 118, 155
parsley roots 119, *140*
parsnips 18–19, 116, 119, 140–141
pasta *tarako* 260
Pastinaca sativa 141
pata negra 155, 238
pâté 262
peaches 23, 119
peanuts 23, 163
pears 50, 58, 73, 112, 141, 248
peas 141. *See also* green peas, yellow split peas
~ aroma 50, 51
~ family. *See* Fabaceae
~ proteins 100
~ soup 188
~ vitamins 116
pectin 8, 58–59, 84, 86–88, 112, 136
~ bonds 59
pectinase 87
Penicillium glaucum 149
Penicillium roqueforti 149
penten-3-one 144
pepper 42, 54, 56, 151, *160*, 164. *See also* kampot pepper, *sansho* pepper
pepper berries 28
peppermint 43
pepperoncini 57
Pepper X 57
peptides 45, 95, 99. *See also* dipeptides, tripeptides
Perilla frutescens 51, 153–154
Persea americana 133
persimmons 42, 58
pesticides 12, 52, 109–110, 125–126
pesto 148, *161*
pests 109, 125
Petroselinum
~ *crispum* 140
~ *crispum* var. *tuberosum* 140
pH
~ Maillard reactions 77
~ texture 38
Phaseolus
~ *coccineus* 133
~ *vulgaris* 133
~ *vulgaris* var. 'Hilda' 133
phenols 50, 76, 115. *See also* polyphenols
phenylalanine 50, 99
phloem 74
phosphorus 12, 118
photosynthesis 8, 30–31, 72–74, 78, 115, 132
phthalides 136
phycobilins 78, 115

phylloxera 64
Physalis spp. 164
physiological taste 42–43
Phytophthora infestans 109
phytoplankton 29, 105
phytosterol 106
pickles 92, 94, 123, 158, *162*, 168. *See also* tsukemono
pickling 72, 83, 92, 121. *See also* conserving, fermenting
~ medium 94, 121, 123, 157
~ melon 94
~ vegetables 92, 167
pigments 43, 58, 71–74, 78, 90, 110, 115
pimenta 56
Pimenta dioica 160
pimiento 56
pineapple 204
pine nuts 161
pink peppercorns 160
piperine 54, 160–161
Piper longum 160
Piper nigrum 160
Piper spp. 160
Pisum sativum 141
~ var. 'Macrocarpon Group' 141
placenta 56
planetary diet 13
plankton 29
Plantae 31
plantains 39
plant breeding 7, 39, 44, 52–53, 119, 126
plant protein 52
plant-rich diet. *See* diet
Pleurotus ostreatus 140
plums 58, 119, 153, 176
~ unripe 59
plum tomatoes 144
Poaceae 137
poblanos 57
poisons 119–120
polenta 256
pollinators 126
pollution 12
Polygonaceae 142
polyphenol oxidase 76
polyphenols 43, 50, 58, 66, 76, 115
polypyrroles 153
polysaccharides 58, 60, 69, 106, 114. *See also* disaccharides, sugar
pomelo 151, 264
ponzu 151–152, 158, *162*, 168, 186
poppy seeds 118
population 8, 11–12, 125
porcini 19, 45, 60, 115, 140, 158

Index 281

pork 77, 102, 104, 148–149, 155, 248
porphyran 112, 114
Porphyra/Pyropia spp. 66, 68–69, 114, 116, 143
Portobello 19, 140
potassium 30, 108, 110, 118
potatoes 13, 18, 19, 44, 108–111, *141*, 198, 248. *See also* sweet potatoes
~ baked 246
~ blight 109
~ browning 76–77
~ chips 77
~ colour 72–73, 110, 119
~ cooked 244
~ cooking water *162*, 165, 190
~ macronutrients 108, 110
~ mashed 58, 232
~ micronutrients 118
~ peel 108, 110
~ purée 242
~ starch 8, 13, 58, 107
~ taste and aroma 45, 50–51, 110, 141
~ texture 88, 141
~ varieties 108–109
~ vitamins 116
poultry 13, 100, 102, 148–149
poverty 11
prebiotics 112
preserving 44, 92, 95, 120–123, 167. *See also* conserving, fermenting, pickling
~ colours 72, 76, 88
~ nutritional value 71
~ texture 72, 84
processed foods 12–13, 15, 99, 113, 120
proteins 47, 58, 87, 99–100, 115. *See also* enzymes, lectins, peptides
~ amino acids 99, 115
~ breakdown 47, 95, 100, 121, 122
~ complete 100
~ contents 30, 100, 110, 123
~ denaturation 6, 29, 99
~ diet 97–100, 102–103, 107
~ folding 99
~ lectins 119
~ Malliard reactions 77
~ meat 6
~ meat substitutes 52
~ plants vs animals 100
~ saliva 58
~ umami 45
~ water binding 58
pseudo-cobalamin 116
psoralens 119
puffballs 60, 140
pulses 28, 112, 118. *See also* beans, legumes, lentils, peas
pumpkin seeds 28, 104, 162, 250
purple beans 71–72, 133
Puy lentils 118, 140
pyrazines 51, 141
Pyrus spp. 141

Q

quinoa 100, 254–255

R

radicchio 58, *142*, 184
radishes 18–19, 54, 72, 83, 87, 116, *142*. *See also* daikon, turnips
~ black 142
ramen 103
Raphanus
~ *sativus* var. *longipinnatus* or var. *acanthiformis* 137
~ *sativus* var. *niger* 142
~ *sativus* var. *sativus* 142
~ spp. 142
ratatouille 206
ravioli 64
raw food 120–121
receptors 54, 56, 122. *See also* taste receptors
red beets 113, 118, 214, 236
red blood cells 116
red cabbage 54, 72, 90, 135
remoulade 158, *162*
Rheum spp. 142
rhizomes 18–19, 38, 90, 138
rhubarb 18–19, 119, *142*
ribonucleotides. *See* nucleotides
ribose 77
rice 8, 13, 92, 102–103, 110, 164, 166
~ bran 92, 123
~ brown 100, 118
~ cooked 34, 36, 69, 156
~ fermented 95, 123, 158, 162
~ paper 194
~ polished 102–103, 107–108, 110
~ vinegar 83, 162, 168
~ wine. *See* mirin, sake
ripening 44–45, 52, 59, 71, 87, 107
risotto 258
rizza 194
roasting 76, 91
roe 2, 149, 260
romaine 83, 142
romanesco 19, 136, 264
Romano beans 133
romesco sauce 20, 22, *163*
roots 7, 17–19, 90
rootstocks 18, 132
root vegetables 18, 58, 108, 119, 230
Roquefort 149
Rosaceae 132, 141, 143
rose family. *See* Rosaceae
roughage 106
rouille 150, *164*
Rousseau, Auguste 64
runner beans 133
Russian kale, red *142*
rutabagas *142*, 208
ryōri 34

S

saccharides 106. *See also* monosaccharides, polysaccharides, sugar
Saccharina
~ *japonica* 66, 143, 150, 152, 165
~ *latissima* 69, 143, 164
Saccharomyces cerevisiae 29, 159
sake 7, 83, 95, *164*
~ *kazu* 123, 157, 164
~ lees 92, 123, 157
salad greens 84, *142*
Salicornia europaea 44
saliva 43, 58, 106, 121–123
salsa verde 164
salt 118. *See also* marinade, salting
salting 121, 148, 152, 167
salty 36, 43, 47
~ enhancement 45, 83
sanbaizu 152, 158, 162
Sanchez, Monica 50
sansho pepper 34, 83, 157, 160, *164*
saponins 132, 145
sashimi 34, 154
satiety 122
sato-zuke 94
saturated fats 104. *See also* fats
sauces 5, 18, 45, 59, 84, 148, 155, 158, 162, 164
~ taxonomy 59
sauce tomate 59
sausages 49, 68
sautéing 76
savoy cabbage 135
scallions 140
scallops 45, 47, 188–189
Schatzker, Mark 52
Schinus molle 160
Scoville Heat Units (SHU) 56
Scoville, Wilbur 56
Scozonera hispanica 134
sea asparagus 18
sea lettuce 19, 34, 68, 78, 83, 143
sea rocket 18
seasonings 1–2, 5, 48, 52, 81–83, 95
seasons 24, 34, 71, 90, 92
seaweeds 1–2, 7, 19, 29, 31, *143*, 188, 254, 266. *See also* bladderwrack, dulse, konbu, macroalgae, *nori*, sugar kelp, tangle, sea lettuce, winged kelp
~ bacon 68
~ colours 78
~ liquorice *164*
~ nutrients 98, 100, 104–107, 112, 114–116, 118
~ powder and granulate *165*
~ taste and aroma 32, 34, 45, 51, 66
~ texture 8, 68–69
~ toxins 120
seeds 7, 13, 17, 26, 28, 34, 38, 56, 58, 83, 84, 120, *162*
~ contents 99–100, 104, 107, 112, 118
selenium 118, 121
senmai-zuke 94
senses 41
sensory perception 5, 41, 56
serranos 57, 151
sesame seeds 28, 34, 83, 107, 116, 118, 153, 155, 162, 166, 186, 214
shallots 140
shellfish 2, 45, 99, 104–105, 120, 149, 152
~ powder *165*
shichi 83
Shichifukujin 83
shichimi 34, 83, 151, 164, *166*, 168
shiitake 19, 29, 32, 51, 60, 116, 140, 158, 214
~ dashi 34
~ umami synergy 45, 152, 159
shio 155
~ -*koji* 95, 156, 178, 180
~ -*zuke* 94
shiro-miso 158
shiro-tosaka 78
shiso 51, 153, 154
shogaol 138, 154
shōjin dashi 32–34, 152
shōjin kaiseki 36
shōjin ryōri 32–35, 83, 94, 158
short-chain fatty acids 113–114
shōyu 34. *See also* soy sauce
shōyu-zuke 94

shrimp heads 165
SHU 56–57
Sichuan pepper 160, 164
Sikhism 32
silver beets 144
Sinapis alba 159
skipjack tuna 156
slimy 69
smell. *See* aroma
smoothie 30
snow peas 141
sodium 30, 45, 88, 110, 118
~ chloride 45, 118
Solanaceae 133–134, 141, 144, 151
solanine 110, 119
Solanum
~ *lycopersicum* 144
~ *melongena* 133
~ *tuberosum* 141
Solera method 148
somatosensory system 43
sound 1, 43, 87, 94
soups 34, 45, 47, 49, 69, 83, 90, 110, 149, 188–190. *See also* bouillon, *dashi*, miso soup
sour 36, 43, 47
soybeans 32, 47, 95, 100, 104, 118, 133, 156
~ fermented 155, 158, 166
~ milk 34
soy sauce 34, 45, 47, 77, 83, 92, 94–95, 157, 162–163, 166, 168. *See also* ponzu, sanbaizu
sp. 132
spears 90–91, 132
species 131
spices 2, 34, 54, 56, 81, 147
spinach 18, 100, 115–116, 118–119, *143*, 192, 202
Spinacia oleracea 143
spirulina 29–30, 116, 120
Spirulina platensis 116
spp. 132
spring onions 20, 24, 140
sprouts 120, 162, 200
squash 28, *143*, 258
~ flowers 28
squid 150, 152
ssp. 132
stalks 17–19, 28, 74, 90
starch 6, 8, 56, 58, 84, 86, 90, 106, 110, 156, 164
~ taste 44, 108
starchy vegetables 13, 58, 119, 141
starvation 12, 109
steaming 41, 58, 110
stems 7–8, 17–19, 44, 68
sterols 105–106

sticky 7, 76
Stilton 149
stocks 12, 45, 47, 59, 90, 149. *See also* bouillon, *dashi*
stomach 6, 112, 114, 123
~ -brain axis 122
strawberries 45, 52, 112, 115, 119, *143*
~ unripe 58
su 162, 168
~ -*zuke* 94
sucrose 77, 86, 106
sugar 87, 90, 95, 106–107, 122. *See also* carbohydrates, fructose, galactose, glucose, saccharides, sucrose
~ caramelisation 76
~ contents 44, 76, 90, 103
~ diet 13
~ Maillard reactions 76–77
~ reduction 47
~ sweetness 44, 83
~ water binding 58, 86
sugar cane 110
sugar kelp 51, 66, 69, 118, 120, 143, 164
sugar snap peas 28, 141
sulfites 76
sulfur compounds 46, 50–51, 54, 60, 66, 115
sulfur dioxide 160
sumac 168
summer truffles 144
sunchokes 138, 232
sunflower oil 104, 116
sunflower seeds 28, 107, 118, 162
superfood 30
supertasters 7
sushi 34, 69, 154
~ factor, the 114
Sustainable Development, Agenda for 11
Suzuki, Saburosuke 48
sweet 6, 36, 43, 106
~ enhancement 45, 47, 83
~ preference 6
sweet potatoes 73, 118, *144*, 250
Swiss chard 18, 19, *144*
symbiosis 56, 62, 74, 114, 116, 144

T

table sugar. *See* sucrose
tactile stimulation 43
tahini *166*
tamarind 46, 163, *166*, 168
Tamarindus indica 166
tangle 51, 66, 118, 120, 164
tangleweed 120

tannins 28, 43, 50. *See also* polyphenols
taproots 13–19
tarako 260
tartaric acid 44
tartar sauce 158
TAS2R38 gene 7
taste. *See also* basic tastes, bitter, salty, sour, sweet, umami
~ appetite 122
~ attribute 45, 47
~ basics of 41–43
~ buds 1, 41–43
~ experience 41–43
~ fungi 29
~ herbs 18
~ microalgae 30
~ mushrooms 60, 100
~ plants 7, 30, 44–47
~ plants vs meat 5–7
~ preferences 2, 5, 13–15, 41, 103. *See also* food preferences
~ receptors 7, 43
~ seaweeds 66
~ vegetables 7, 18, 54, 99
~ vs aroma 50
taurine 98
tea 36, 43, 49, 119
Tenochtitlán 30
temperature sensing 43, 54
temple cuisine 1, 34, 36, 83, 94, 158
tempura 150, 160
terpenes 50–51, 133–135, 138
terpenoids 140
Texcoco, Lake 30
texture 1–2, 7–8, 42–43, 49, 81, 84, 88, 92
thickening 8, 59, 84, 112, 150, 164
thiocyanates 142
threonine 99
thyme 18, 50, 263
thymol 50
tikka masala 59
toasty 76
tocopherol 115
tofu 34, 118, 184
tomatillos 164
tomatl 59
tomatoes 19, 28, 52–53, *144*, 212
~ colour 73
~ ketchup 156
~ pulp 45, 59
~ purée 166
~ sauces 59, 218
~ sun-dried and in paste *166*
~ sun-ripened 45, 144, 166
~ taste and aroma 45, 51–52

~ texture 88, 268
~ umami 45
~ vitamins 116
tonka beans 119
tortillas 52, 198
tosaka-nori 78
toxins 7, 110, 112, 115, 119–120, 122, 125, 133
trace elements 118, 120–121
trigeminal sensation 42–43, 151
tripeptides 47, 60, 138, 147, 153
triterpenes 137
trufficulture 62, 64
truffles 19, 51, 60, 62–65, *144*, 246
tryptophan 99
tsukemono 92–94, 123, 133, 137–138, 142, 156, 162, *167*, 172–173, 180
~ marinade 167
~ nutrients 123
Tuber
~ *aestivum* 144
~ *magnatum* 144
~ *melanosporum* 144
~ spp. 144
~ *terrae* 62
Tuberaceae 144
tubers 8, 18–19, 108–110, 118–119
turgor 44, 58, 87, 91
turmeric 18–19, 138
turnips 19, 26, 58, 94, *144*
Tuscany kale 139

U

Ueda, Yoichi 47
ultra-processed food 52, 128
Ulva lactuca 68, 143
Ulva linza 143
Ulva spp. 78
umai 49
umami 1, 36, 48, 83, 122. *See also dashi*, glutamate, nucleotides
~ basal 45, 60
~ basic taste 43
~ discovery 48
~ enhancing salty and sweet 83
~ garlic 46
~ *koku* 45–47
~ meat 6
~ mushrooms 29, 32, 60, 65
~ potatoes 110
~ preference 6
~ reducing bitterness 83
~ reducing fat 110
~ ripening 45

~ seaweeds 66
~ synergistic 45, 47, 122
~ synergy 45, 60, 148–149, 152, 158, 165
~ tomatoes 45, 59
~ vegetables 7, 45, 92, 94, 95
Umami Science Square 48
umbellifer family. See Apiaceae
umeboshi 153
umesu 153
Undaria pinnatifida 66, 68, 78, 105, 143
United Nations 11, 13
unsaturated fats 104. See also fats, omega-3, omega-6
uri nare-zuke 94

V

vacuoles 50, 86–87
valine 99
vanillin 50
var. 132
variety 132
vascular tissue 74, 90
veal 149
vegan 1–2, 15, 32, 34, 60, 83, 98, 100, 115–116, 120, 127–129, 158, 171
~ cuisine 159. See also *shōjin ryōri*
~ *dashi* 152
Vegemite 159
vegetabilis 17
vegetables 5–7, 13–15, 17–19, 84, 99, 116
~ blanching 72
~ colours 71–74
~ cooked 44, 72, 107
~ cooking 88, 119
~ nutrients 98, 100, 104, 107, 112, 115, 118–119
~ poisons 119
~ preferences 41–42
~ preserving and fermenting 121–123
~ raw 120–121
~ sauces 59
~ taste and aroma 7, 29, 44, 50–51, 54, 58
~ texture 7–8, 59, 84–88
~ umami 7, 45–46
vegetarian 7, 15, 32, 34, 94, 98, 116, 127, 129, 152, 159, 171. See also vegan
velouté 59
verjus 167
Vicia faba 133
Vigna mungo 140
Vigna radiata 140
vinaigrette 148, *168*
vincotto 168
vinegar 68, 72, 88, 152, *168*. See also balsamic vinegar, rice vinegar
~ gastriques 154
~ mayonnaise 158
~ mustard 159
~ pickling 92, 94, 121, 162, 167
~ vinaigrette 168
vitamin
~ A 73, 115, 123
~ B 110, 115, 123
~ B_{12} 98, 115–116, 121
~ C 76, 110, 115, 123
~ D 98, 105, 121
~ E 115
~ K 115
vitamins 24, 29, 95, 97–99, 115, 121–123
von Liebig, Justus 109

W

wakame 66, 68–69, 78, 105, 116, 143, 184
Waldorf salad 136
walnuts 45, 104, 107, 156
wasabi 19, 54
waste. See food waste
water binding 8, 58, 86, 112
water content 44, 53, 84, 86–87, 94, 123
watermelons 73, 140, 270
Western diet 5, 103, 105, 108, 114
wheat 13, 34, 103, 107, 110, 118, 160, 166
~ germ 118
white beans 88, 118, 266
white cabbage 135
white pepper 160
white truffles 144
whole foods 15, 52, 129
whole grains 13, 107, 118. See also grains
wild sea beet 144
wine 7, 43, 50, 76–77, 157–158, 168
~ vinegar 121
winged kelp 78, 143
winter truffles 144
witloof 134
wood ear mushrooms 192
wood sorrel 119
woody 7–8, 44, 60, 90–91
Worcestershire sauce 166, *168*
world population 8, 11–12, 125
Wrangham, Richard 6

X

xitomatl 59
xylan 112
xylem 38, 74, 136

Y

yams 18, 73, 144–145
yeast 29, 47, 157
~ nutritional/flakes *159*, 208
yellow split peas 100, 118, 263
yogurt 2, 116, 192, 240
Yoshimura, Hiroyuki 116
yuba 34
yukari 153
yuzu 34, 151, *168*
~ peel 83, 153, 166

Z

za'atar 168
Zanthoxylum
~ *piperitum* 160, 164
~ *simulans* 160
Zea mays 137
zeaxanthin 110
Zen 1, 32–36, 60, 103
zinc 108, 113, 118
zingerone 138, 154
Zingiberaceae 138
zingiberene 138, 154
Zingiber officinale 138, 154
zucchinis *145*, 218, 228
~ flowers 28

Acknowledgements

A number of colleagues and good contacts have been very helpful to us when working on this book. We wish to acknowledge the following: Kasper Styrbæk for aid and assistance when developing recipes and preparing for photo sessions; Britt Nilsson and Christoffer Huus for testing recipes and for constructive suggestions for improvements; members of the national Danish centre Taste for Life for inspiring work on taste; Motonaka Kuroda for information about *koku* and *kokumi* in Japanese cuisine; Mieko Yoshida for guidance and interpretation during a visit to the Kanga-an temple (*shōjin ryōri* meal) and the Higuchi-farm, Kyoto, as well as a reference to the Japanese mnemonic device *mago-wa-yas-ashii*; Masataka Higuchi, fourteenth-generation owner of the Higuchi-farm, for information about the farming of Kyoto vegetables (*kyō-yasai*); Lars Andersen (Knabegaarden), Peter Bay Knudsen and Marie Ejlersen (Skiftekær), as well as Anders Christian Mortensen (Holistia) for access to, and guided tours of, their respective farms.